·Andrew McLeish·Ron Grigson·

LHBEC

GEOLOGICAL SCIENCE
QUESTIONS AND ANSWERS

Blackie

GEOLOGICAL SCIENCE
QUESTIONS AND ANSWERS

ISBN 0 216 92529 0
First published 1989
© Andrew McLeish and Ron Grigson 1989

Illustrated by Andrew McLeish

Published by Blackie & Son Ltd
Bishopbriggs, Glasgow G64 2NZ
7 Leicester Place, London WC2H 7BP

British Library Cataloguing in Publication Data

McLeish, Andrew
  Geological science: questions and answers.
  1. Geology—For schools
  I. Title
  551

ISBN 0-216-92529-0

Filmset by Advanced Filmsetters (Glasgow) Ltd
Printed in Great Britain by Bell and Bain Ltd., Glasgow

This book has been written for students studying A-level and H-grade courses in geology. It will also be of value to those in the early stages of college and university courses. The questions cover all of the major syllabus topics and give full cognizance to the many applications of geology. Most questions are of structured type and many are based on illustrations and data from real situations. The questions have been pitched at various levels of difficulty; they will help improve students' levels of knowledge and understanding, and their abilities to interpret information and to solve geological problems.

The answers are clear and concise, providing all the required detail and information. At times, it may be possible to interpret questions in ways outwith those given in the answers, so there will be scope for extended classroom discussion. Finally, since the work will have its imperfections, we would be pleased to hear of questions and answers which are, respectively, ambiguous or incorrect.

Andrew McLeish

Ron Grigson died before this book was completed. He provided the initial inspiration for a project which has subsequently been brought to truition by Andrew McLeish as **Geological Science—Questions and Answers**.

Preface

The authors and publishers would like to express their gratitude to the following:

The Director of the Scottish Examination Board for permission to reproduce questions 3.6 (a) (i)–(iii); 3.7; 4.5 (a), (b), (c); 4.10 (d); parts of 4.11 (b), (c); 5.3 (a); 5.6; 5.7; 5.8; 5.9; part of 6.2 (a); 7.3; 8.2; 8.3; 8.4 (a), (c), (d); 8.5; 8.6; 8.15; 10.6 (a), (b), (g), (i), (j), (l), (n); 12.6 (a); and 12.11
The great majority of these questions had previously been written by one of the authors for use in geology examinations set by the Scottish Examination Board. It should be noted that the answers given to these questions have not emanated from the Scottish Examination Board.

The editors of the *Scottish Journal of Geology* for permission to quote information in questions 4.4; 6.7; 8.1; 11.2; 12.7; and 12.12.

The Director of the British Geological Survey for permission to reproduce Figures 12.9A and 12.9B. (NERC copyright reserved.)

Professor A. G. Kemp for permission to use information and diagrams in question 12.18 (c).

Dr C. Gillen for permission to use questions 4.11 (c); 7.3; and 8.2.

The Director of British Coal for permission to reproduce Figure 12.14B: (Taken from *Subsidence Engineers' Handbook*—1975)

Mr A. A. McMillan for advice on question 12.14.

The following are thanked for permission to use copyright photographs:
Aerofilms: Questions 3.11 and 8.12
Dr C. Gillen: Questions 7.3 and 8.2
National Remote Sensing Centre: back cover photograph
ERSAC Ltd: Question 12.17

The front cover photograph was kindly supplied by British Petroleum Ltd

The following diagrams have appeared in previous publications: Figs. 4.11B and 4.11C—after Goudie, Cooke and Evans (1970). 'Experimental investigation of rock weathering by salts'. *Area* **4** pp. 42–48.
Fig. 3.10C—from *Earthquake Information Bulletin* (1984) **16**, No. 2. U.S. Department of the Interior, Geological Survey.

Acknowledgments

Contents

PLANET EARTH

1.1 The solar system may have formed from a nebula of dust and gas. It is thought that the events shown in Table 1.1 occurred as the nebula cooled.

Temperature (°C)	Event
1300	Condensation of oxides such as CaO and Al_2O_3.
1000	Condensation of Fe–Ni alloy.
900	Condensation of pyroxene.
	From 900–200 °C, Fe is oxidized to FeO which reacts with pyroxene to form olivine.
700	Reaction of Na with Al_2O_3 and silicates to form feldspar.
400	Reaction of H_2S with Fe to make FeS.
300	Reaction of H_2O with Ca minerals to make amphibole.
150	H_2O reacts with olivine to make serpentine.
−100	Condensation of ice.
−120	Ammonia reacts with ice to make solid $NH_3.H_2O$.
−200	Condensation of Ar and methane to solid forms.
−250	Condensation of Ne and H.
−270	Condensation of He.

Table 1.1

(a) In which part of the solar nebula would the temperature have been highest? Which types of material would have condensed in this area?

(b) On what two factors does the density of a planet depend?

(c) Why do Mercury, Venus, Earth and Mars have much greater densities than the other planets?

(d) Explain why Mercury has a very high density for its size.

(e) Why would light gaseous material tend to move to outer parts of the nebula?

(f) During accretion, what would happen to the gravitational force of attraction between masses as their sizes increased? How would this affect the impact velocities of infalling particles?

(g) Into what form of energy would the kinetic energy of infalling particles be mostly converted?

(h) Heat produced by accretion and by short-lived radioisotopes would melt a planet or planetesimal (minor planet) and cause iron to be released from silicates:

$$\text{iron olivine} \longrightarrow \text{iron pyroxene} + \text{iron oxide}$$
$$\text{iron oxide} + \text{carbon} \longrightarrow \text{iron} + \text{carbon dioxide}$$

What would be the source of the carbon? What would happen to the liquid metallic iron?

1.2 (a) How can age relationships among lunar structures be established?

(b) Place the structures in Fig. 1.2 in order of formation from old to young.

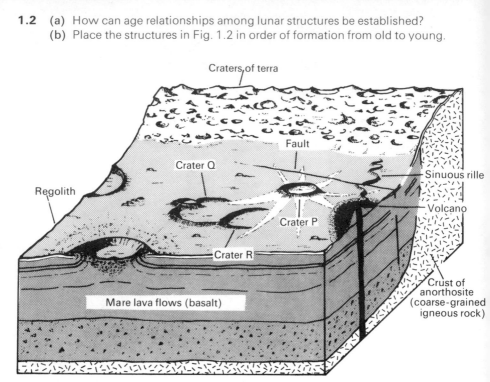

Craters of terra

Fault

Crater Q

Sinuous rille

Regolith

Volcano

Crater P

Crater R

Crust of anorthosite (coarse-grained igneous rock)

Mare lava flows (basalt)

Fig. 1.2

(c) How can study of crater densities be used to establish the relative ages of different areas?

(d) What are crater rays and how are they formed?

(e) Why do old craters lack rays and sharply-defined rims?

(f) How has the lunar regolith been formed?

(g) On the Moon, where might older material overlie younger material?

(h) Sinuous rilles are thought to be collapsed lava tubes. How are lava tubes formed?

(i) Suggest two ways in which a chain of small craters may be formed.

1.3 Table 1.3 shows the areas of land and sea floor at various heights and depths.

(a) What is the total land area? What is the total area of the sea floor?

(b) Complete the table to show each area as a percentage of the total area of land and sea floor.

(c) Draw a histogram to show the percentage areas of heights and depths. (Plot height or depth vertically with the highest areas on the left and the deepest on the right.)

(d) What are the most common levels of land and sea floor? Explain why these distinct levels suggest that continental and oceanic crust are fundamentally different.

Height (m)	Land surface area ($km^2 \times 10^6$)	% of total land and sea floor area	Depth (m)	Sea floor area ($km^2 \times 10^6$)	% of total land and sea floor area
0–1000	106.6		0–1000	18·2	
1000–2000	22.6		1000–2000	15.2	
2000–3000	11.2		2000–3000	24.5	
3000–4000	5.8		3000–4000	70.8	
4000–5000	2.2		4000–5000	119.1	
above 5000	0.5		5000–6000	81.1	
			6000–7000	4.0	
			below 7000	0.4	

Table 1.3

1.4 (a) Identify the numbered features in Fig. 1.4.

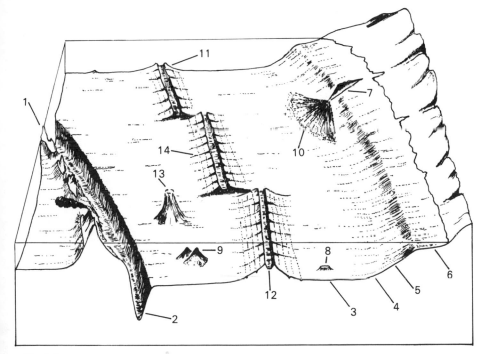

Fig. 1.4

(b) What are cratons and platforms? Give four locations of each feature.
(c) What is an orogenic belt? Name four young mountain ranges (less than 300 Ma old) and four ancient mountain ranges (600–300 Ma old).

MINERALS

2.1 (a) Table 2.1A shows the radii of ions commonly found in minerals.

Ion	Radius (*picometres*)
Na^+	97
Ca^{2+}	99
Mg^{2+}	66
Fe^{2+}	74
Al^{3+}	51
Si^{4+}	42
O^{2-}	140

Table 2.1A

Table 2.1B shows relationships between radius ratio and coordination number for negative ions around positive ions.

Ratio of radius of positive ion to radius of negative ion	Arrangement of negative ions around positive ion	Coordination number (Number of negative ions around positive ion)
0.22–0.41	At corners of tetrahedron	4
0.41–0.73	At corners of octahedron	6

Table 2.1B

(i) How does an electrovalent chemical bond differ from a covalent bond?

(ii) What is the coordination number of an ion?

(iii) What are the ratios of the radii of the following ions?
1 Si^{4+}/O^{2-} 2 Al^{3+}/O^{2-} 3 Mg^{2+}/O^{2-}
What are the expected coordination numbers of O^{2-} ions around Si^{4+}, Al^{3+} and Mg^{2+}?

(iv) Account for the following observations.
1 High temperatures and low pressures favour low coordination numbers. Low temperatures and high pressures favour high coordination numbers.
2 In minerals formed at high temperature or low pressure, Al^{3+} is usually surrounded by four O^{2-} ions. In minerals formed at low temperature or high pressure, Al^{3+} usually occurs in six-fold coordination with O^{2-}.

(b) (i) What are isomorphous or isostructural substances?
(ii) What is solid solution? Explain why isomorphism is not the same as solid solution.
(iii) What are polymorphs?
(iv) What are phases, phase changes and phase diagrams? How many phases are present in a solution of NaCl?
(v) What is a eutectic mixture?
(vi) Account for the following observations.
    1 The olivines Mg_2SiO_4 and Fe_2SiO_4 are both isomorphous and capable of complete solid solution.
    2 When Al^{3+} substitutes for Si^{4+} in feldspar, Ca^{2+} also substitutes for Na^+.

2.2 **(a)** (i) What is the relative density (specific gravity) of a substance?
(ii) When finding the relative density of a mineral why must the specimen be pure?
(b) Why do minerals such as olivine have a range of relative densities?

2.3 **(a)** How does cleavage differ from fracture? Name one mineral which shows cleavage of the types shown in Fig. 2.3A.

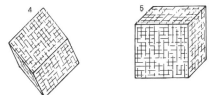

Fig. 2.3A

(b) What is a twinned crystal? Which minerals show twinning of the types shown in Fig. 2.3B?

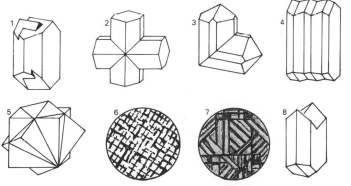

Fig. 2.3B   6 and 7 show twinning as it would appear in thin section under crossed polarized light.

(c) Identify the following minerals from the descriptions.

1 On heating, goes black and gives off water and carbon dioxide. Colours a flame green. Pale green streak. Often forms botryoidal or reniform masses. Typically found in oxidation zone of mineral deposits formed by secondary enrichment.

2 Often forms tetrahedral crystals. Good cleavage in six directions. Gives off hydrogen sulphide with hydrochloric acid. Colour usually brown or black. Resinous or adamantine lustre. Found in hydrothermal and metasomatic mineral deposits.

3 Relative density 5. No cleavage. Often found as reniform masses. Forms black, shiny rhombohedral crystals. Becomes magnetic on heating. Common accessory mineral of igneous rocks. Common ore of sedimentary origin.

4 Forms hexagonal crystals. Lustre vitreous or subresinous. White streak. Colour usually pale green, blue-green or yellow-green. Hardness 5. Used to make phosphate fertilizer.

5 Gives off carbon dioxide with hot hydrochloric acid. Forms rhombohedral crystals with curved faces. Three perfect cleavages. Colourless, white or honey-coloured. Often formed diagenetically in limestones.

6 Does not form crystals. Occurs as massive, granular or fibrous forms. Colour variable (green, black, red, yellow, brown). Often mottled and veined. Cleavage not always visible. Lustre subresinous or greasy. May feel slightly soapy. Often formed by the alteration of ultrabasic igneous rocks.

7 Usually massive. Brassy yellow, often with an iridescent tarnish. Greenish black streak. Turns a flame green. Soluble in nitric acid. Hardness $3\frac{1}{2}$–4. Often found in hydrothermal deposits.

8 Often forms long crystals with triangular sections and longitudinal striations. Colour may change when viewed from different directions. Hardness 7. Found in pegmatites and in granite acted on by boron-rich fluids.

9 Usually white. May form cockscomb crystals. Cleavage in three directions. Relative density 4.5. Colours a flame yellow-green. Common gangue mineral. Also deposited by hot springs.

10 Usually massive. Shiny black streak. One perfect cleavage. Greasy feel. Good conductor of heat. Hardness $1\frac{1}{2}$. Found mostly in metamorphic rocks.

2.4 Describe the arrangement of the SiO_4 tetrahedra in the following groups of minerals.

(a) olivine, garnet   (b) pyroxenes   (c) amphiboles   (d) talc, mica, clay
(e) quartz, feldspar

For each mineral group, state briefly how the physical properties are related to the atomic structure.

2.5 (a) The phase diagram of the system diopside (pyroxene)—anorthite (calcium plagioclase) is shown in Fig. 2.5A.
   (i) At what temperatures do anorthite and diopside melt?

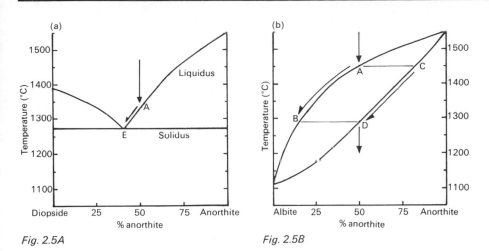

Fig. 2.5A   Fig. 2.5B

(ii) At what temperature would a melt of composition anorthite 50%—diopside 50% begin to crystallize? Which mineral would crystallize first from this melt?

(iii) After meeting the liquidus, the composition of the melt changes along A—E. Which melt composition and temperature are represented by point E? What happens to the melt at point E?

(iv) Describe the behaviour of these melts as they cool:
1 anorthite 42%—diopside 58%
2 anorthite 10%—diopside 90%

(b) The phase diagram of the anorthite—albite system is shown in Fig. 2.5B. Albite is sodium plagioclase. Anorthite and albite show complete solid solution.

(i) At what temperature will a melt of composition anorthite 50%—albite 50% begin to crystallize? At what temperature will crystallization be complete?

(ii) Solid of composition C is first to crystallize. What is the composition of solid C?

(iii) The solid reacts continuously with the liquid so the composition of the solid changes along C—D. What is the composition of the solid which finally crystallizes?

(iv) As the solid changes composition along C—D, the liquid changes composition along A—B. What is the composition of the last liquid to crystallize?

(v) Zoned feldspars such as that shown in Fig. 2.5C are often found in igneous rocks. The crystal has formed from a melt consisting of anorthite 60%—albite 40%. How does such zoning develop?

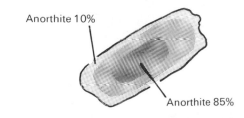

Fig. 2.5C

Anorthite 10%

Anorthite 85%

2.6 **(a)** (i) What is a crystal?

(ii) What is a grain?

(iii) What is a crystal lattice?

(b) (i) Explain the meaning of: plane of symmetry; axis of symmetry; centre of symmetry.

(ii) Why do crystals have no axes of 5-fold symmetry?

(c) (i) In crystals, how does habit differ from form?

(ii) Fig. 2.6 shows calcite crystals of three types. Which crystal habits are illustrated? Which forms are present in each crystal?

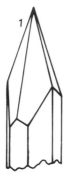
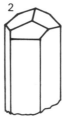
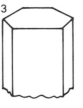

Fig. 2.6

(iii) To which crystal system does calcite belong? Which symmetry elements are possessed by this crystal system?

(iv) Which symmetry elements are diagnostic of the other crystal systems?

2.7 **(a)** (i) What is refraction?

(ii) What is the refractive index of a transparent material?

(iii) Fig. 2.7A shows a light beam being refracted on passing into a mineral. The angle of incidence is 30°; the angle of refraction is 19°. What is the refractive index of the mineral?

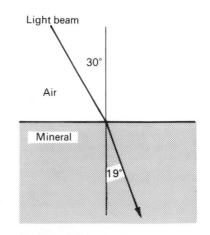

Fig. 2.7A

(iv) Thin sections of rocks are mounted in a medium (e.g. Canada balsam or resin) with a refractive index (RI) of 1.54. Minerals in a rock have the following refractive indices:

Mineral 1: RI = 1.545; *mineral 2*: RI = 1.480; *mineral 3*: RI = 1.740.

Describe the degrees and types of relief shown by the minerals.

(v) When a dot on a piece of paper is viewed through a cleavage rhomb of Iceland spar (transparent calcite), two dots are seen. How is the double image produced?

(vi) How does an isotropic mineral differ from an anisotropic mineral?

(vii) How are the polarization colours of minerals produced by a petrological microscope?

(viii) In a rock consisting entirely of quartz, why do some grains remain black under crossed polars as the stage of the microscope is rotated? Why do other grains change from white to black four times during a rotation?

(b) Place the following minerals in the key of optical properties shown in Fig. 2.7B. Analcite, andalusite, augite (pyroxene), biotite, calcite, chlorite, hornblende (amphibole), leucite, muscovite, olivine, orthoclase, plagioclase, quartz.

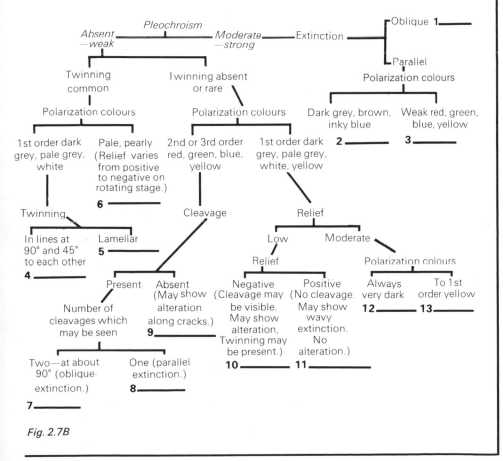

Fig. 2.7B

IGNEOUS ROCKS

3.1 The columns in Fig. 3.1 show the approximate mineral compositions of igneous rocks.

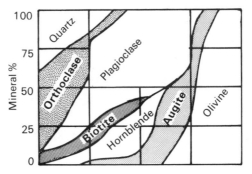

Fig. 3.1

(a) (i) In igneous rocks, how do essential minerals differ from accessory minerals?
(ii) What are the essential minerals of dolerite and rhyolite?
(b) (i) A rock consists of hornblende and plagioclase with smaller amounts of orthoclase, biotite and augite. To which group of igneous rocks does it belong? Name six rocks in this group.
(ii) Name an ultrabasic (ultramafic) rock.

3.2 Bowen's reaction series (Fig. 3.2) shows the sequence of crystallization from a basic magma.

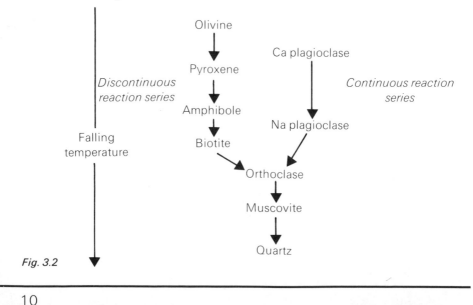

Fig. 3.2

The sequence on the left is a discontinuous series. As the magma cools, olivine crystallizes. With further cooling the olivine reacts with the magma to form pyroxene. The pyroxene may then react with the magma to form amphibole and the amphibole may react to form biotite. The sequence on the right is a continuous reaction series. Ca plagioclase crystallizes first, With cooling, the early-formed plagioclase reacts with the magma to form a progressively more sodic plagioclase. The sequence at the bottom merely shows the order of crystallization. Orthoclase, muscovite and quartz do not react with each other.

(a) By reference to Bowen's series explain the following observations,

    (i) Igneous rocks may show corona texture in which crystals of olivine have augite rims and crystals of augite have hornblende rims.

    (ii) Rhyolite may contain phenocrysts of iron olivine. Granite does not contain olivine.

(b) How will removal of early-formed crystals affect the composition of the remaining magma?

3.3 Describe the events which led to the formation of the igneous textures in Fig. 3.3.

(a)   (b)   (c)

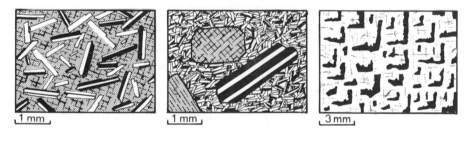

(d)   (e)   (f)

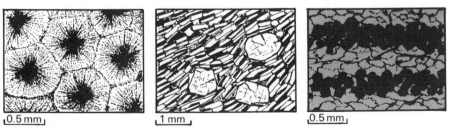

Fig. 3.3

3.4 Triangular diagrams are often used in geology to show relationships among three variables. In Fig. 3.4 the proportion of A decreases from 100% A at the apex of the triangle to 0% A at the base of the triangle. The percentages of B and C similarly decrease away from their respective corners. The proportions of A, B and C at points in the triangle can be read from the percentage grid lines. For example, if A, B and C were three minerals, the mineral composition represented by point P in Fig. 3.4 would be: A 70%; B 30% and C 0%.

(a) What are the compositions represented by points Q and R in Fig. 3.4?

(b) In Fig. 3.4, name the coarse-grained igneous rocks which have mineral compositions represented by points U–Z.

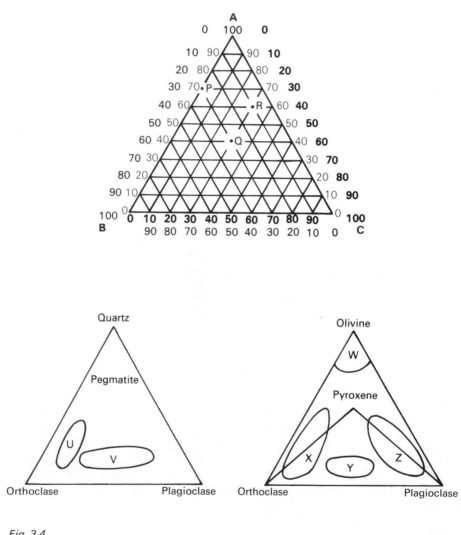

Fig. 3.4

3.5 **(a)** Table 3.5A shows the weight percentages of the main oxides in oceanic crust basalt and in unmelted mantle peridotite.

Oxide	Basalt	Peridotite
SiO_2	49.2	44.5
TiO_2	1.4	0.1
Al_2O_3	15.8	3.1
Fe_2O_3	2.2	1.2
FeO	7.?	6.7
MgO	8.5	39.1
CaO	11.1	3.2
Na_2O	2.7	0.25
K_2O	0.26	0.04

Table 3.5A

It is thought that oceanic basalts are produced by the partial melting of mantle peridotite. What are the main compositional differences between the partial melt (represented by the basalt) and the original rock (represented by the peridotite)?

(b) A basaltic magma has the following composition (oxides in weight per cent):

SiO_2	49.0	MgO	7.6
TiO_2	1.0	CaO	11.0
Al_2O_3	18.2	Na_2O	2.5
Fe_2O_3	3.2	K_2O	0.9
FeO	6.0	H_2O	0.4

(i) If 5 tonnes of magnesium olivine (Mg_2SiO_4) and 5 tonnes of calcium plagioclase ($CaAl_2Si_2O_8$) crystallized and sank from 100 tonnes of magma, what would be the SiO_2 percentage of the remaining liquid?

(ii) Why does the SiO_2 percentage increase?

(c) The compositions of island arc basalt, andesite and rhyolite are shown in Table 3.5B. (The oxides are given in weight per cent.) It is thought that the lavas may represent a sequence produced by partial melting and magmatic differentiation, as shown in Fig. 3.5A overleaf.

Oxide	Basalt	Andesite	Rhyolite
SiO_2	48.3	59.1	74.2
TiO_2	0.6	0.6	0.3
Al_2O_3	19.1	17.1	13.3
Fe_2O_3	2.8	2.4	0.9
FeO	7.0	5.0	0.9
MgO	8.0	2.7	0.3
CaO	11.2	5.7	1.6
Na_2O	2.0	3.1	4.2
K_2O	0.2	2.6	3.2

Table 3.5B

Water from descending oceanic crust of subducted plate lowers melting point of overlying peridotite ⟶ Partial melting of mantle above Benioff zone ⟶ Basaltic magma

Magmatic differentiation and partial melting of diorite ⟵ Andesitic magma ⟵ Magmatic differentiation and partial melting of gabbro

Rhyolitic magma

Fig. 3.5A

What are the main trends in chemical composition shown by the basalt→andesite→rhyolite sequence?

(d) Fig. 3.5B is a plot of K_2O percentage in andesites from an island arc against depth to the centre of the Benioff zone.

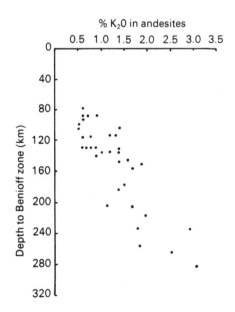

Fig. 3.5B

What general relationship is shown by the graph? Account for this relationship.

3.6 **(a)** Fig. 3.6 shows a vertical section through a sill. Table 3.6A gives information on the minerals in the rock.
   (i) Explain why Zones I and IV are finer-grained than Zones II and III.
   (ii) What is a cumulate rock?
   (iii) Explain why Zone II has 5% magnetite while Zone III has no magnetite.

(iv) If you wanted to collect a rock sample most nearly representative of the whole sill, from which zone would you collect it? Explain your answer.

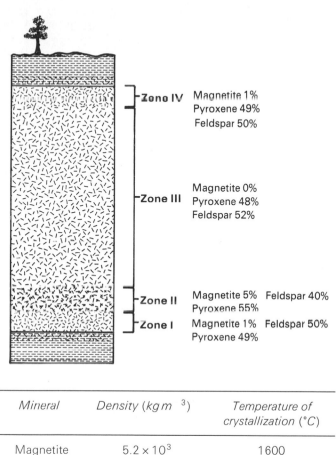

Fig. 3.6

Mineral	Density ($kg\,m^{-3}$)	Temperature of crystallization (°C)
Magnetite	5.2×10^3	1600
Pyroxene	3.28×10^3	1250
Feldspar	2.68×10^3	1250

Table 3.6A

(b) The rate at which a spherical crystal sinks through a magma is given by the equation:

$$\text{settling velocity} = \frac{2gr^2}{9\eta}(\rho_s - \rho_m)$$

where: g is the acceleration due to gravity ($9.1\ \mathrm{m\,s^{-2}}$);
r is the radius of the crystal;
ρ_s (*Greek letter rho*) is the density of the crystal and ρ_m is the density of the magma;
η (*Greek eta*) is the dynamic viscosity of the magma in newton seconds per square metre.

(i) The density of a basic magma is $2.58 \times 10^3 \, kg \, m^{-3}$ and its dynamic viscosity is $300 \, N \, s \, m^{-2}$. Complete Table 3.6B to show the velocities in metres per year at which the crystals sink.

Mineral	Plagioclase		Olivine		Pyroxene		Magnetite	
Density $(kg \, m^{-3})$	2.68×10^3		3.70×10^3		3.28×10^3		5.20×10^3	
$\rho_s - \rho_m$								
Crystal radius $(m \times 10^{-3})$	0.5	1.0	0.5	1.0	0.5	1.0	0.5	1.0
Settling velocity $(m \, yr^{-1})$								

Table 3.6B

(ii) A magma chamber has a vertical height of 900 m. How long would it take the crystals in part (i) to sink if they formed at the roof of the chamber?

(iii) What processes in the magma chamber might alter rates of crystal settlement?

3.7 A sequence of lavas was analysed and the results are shown in Table 3.7. It is thought that the lavas came from more than one volcano.

	Number of lava flow	Sodium oxide + potassium oxide (weight %)	Silicon dioxide (weight %)
Young ↑	12	4.7	50.0
	11	8.8	59.5
	10	8.5	57.5
	9	4.4	47.5
	8	4.1	44.0
	7	5.9	58.5
	6	8.0	55.0
	5	5.3	54.5
	4	5.0	52.5
	3	6.9	50.0
	2	5.9	45.5
Old ↓	1	5.1	41.5

Table 3.7

(a) Plot the percentage weight of sodium oxide + potassium oxide against the percentage weight of silicon dioxide for the twelve lavas. Write the numbers of the lava flows beside the points plotted.

(b) From how many volcanoes did the lavas come? Give a reason for your answer.

(c) What is the general relationship between the amount of silicon dioxide and the amount of sodium oxide + potassium oxide?

(d) Select the **three** correct statements from the following list.

A The amount of silicon dioxide in the lavas increases steadily with time.

B The ratio of sodium oxide + potassium oxide to silicon dioxide increases steadily with time.

C The amount of silicon dioxide in the lavas does not increase steadily with time

D The ratio of sodium oxide + potassium oxide to silicon dioxide is approximately the same for all lavas.

E The ratios of sodium oxide + potassium oxide to silicon dioxide can be used to divide the lavas into groups.

F The volcano which produced lava 1 also produced lava 11.

G Lavas 3 and 12 came from the same volcano.

H The volcano which produced lava 1 also produced lava 7.

3.8 A volcanic cone has a height of 2 km and a diameter of 6 km.

(a) What is the volume of the cone?

After a very violent eruption in which the cone was blown away, ash and other pyroclastic material covered an elliptical area 20 km by 8 km to an average thickness of 100 m. 1.2 km^3 of fine ash were lost into the atmosphere.

(b) What is the total volume of ash which became airborne as a result of the eruption?

(c) What is the volume of ash not accounted for?

(d) What became of this material?

(e) What name is given to the large basin-like crater produced when a volcanic cone collapses into a magma chamber or when a cone is blown away?

3.9 Table 3.9 shows the amounts of energy liberated by some of the world's largest eruptions.

Volcano	Year	Energy (joules)
Tambora	1815	8.4×10^{19}
Sakarajima	1914	4.6×10^{18}
Krakatoa	1883	1.0×10^{18}
Fuji	1707	7.1×10^{17}
Capelinhos	1957	4.0×10^{17}

Table 3.9

The amount of energy given off by Krakatoa is about equal to that from 50 000 early hydrogen bombs.

What information would you need before you could find the amount of energy given off during an eruption?

3.10 Since the violent eruptions of 1980, the activity of Mount St Helens, USA, has been closely monitored by measurements of tilt and of lava dome and thrust fault movements; by seismic studies; and by gas emission studies.

(a) *Tilt measurements*

Fig. 3.10A shows changes in radial tilt on the volcano during the early part of 1982.

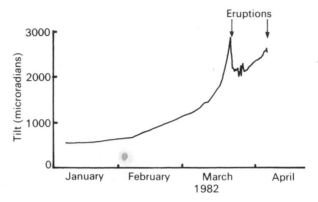

Fig. 3.10A

What happens to the rate of tilting a few days before the eruption?
What happens minutes or hours before an eruption?
Explain both effects.

(b) *Lava dome movements*

(i) The lava from Mount St Helens is dacite. What is the coarse-grained equivalent of dacite?

(ii) Why does the lava form domes rather than lava flows?

(iii) During April and May, 1982, the distance between a target on the dome and a survey station was changing as shown in Table 3.10A. The movement preceded an eruption. Draw a graph of reducing distance against time. Every time you plot a point on the graph, say when you think the eruption will occur.

What happens to the rate of change of distance before an eruption? Why does this happen?

Date, 1982	Reduction in distance between target and survey station (m)
April 19	0.00
20	0.01
21	0.02
24	0.04
26	0.08
30	0.10
May 5	0.24
6	0.26
11	0.79
12	1.13
13	1.95

Table 3.10A

(c) *Movements on thrust faults*

Distances between fixed points on opposite sides of thrust faults on the crater floor are measured as shown in Fig. 3.10B.

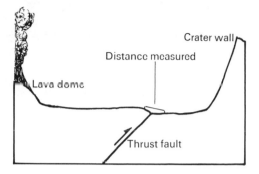

Fig. 3.10B

(i) Why do thrust faults develop on the crater floor?

(ii) Changing distance across a thrust fault during part of 1981 is shown in Table 3.10B. Draw a graph of distance reduction against time. As before, every time you plot a point on the graph, say when you think an eruption will occur.

What happens to the rate of change of distance just before an eruption? Why does this happen?

Date, 1981		Reduction in distance across thrust fault (m)
June	24	0.00
July	7	0.05
August	2	0.15
	15	0.22
	28	0.50
September	1	0.75
	6	1.75
	10	1.80
	23	1.95
October	6	2.00
	19	2.25
	25	2.70
	30	3.20
November	1	3.20
	14	3.20

Table 3.10B

(d) *Seismic studies*

The main types of seismogram recorded by seismometers near Mount St Helens are shown in Fig. 3.10C.

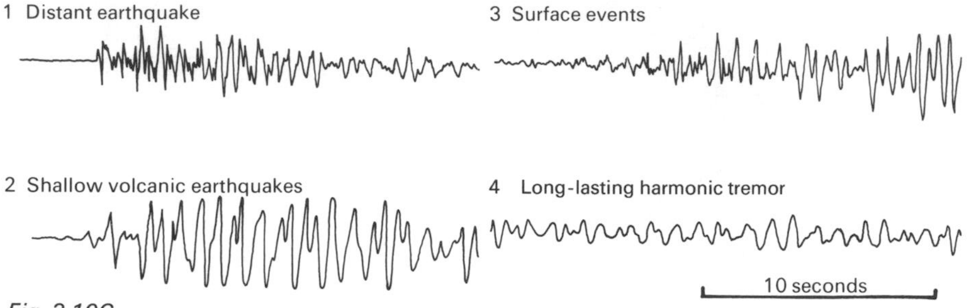

1 Distant earthquake

3 Surface events

2 Shallow volcanic earthquakes

4 Long-lasting harmonic tremor

10 seconds

Fig. 3.10C

(i) What kinds of movements have produced these seismograms?
(ii) Account for the following observations:
    1 An increased number of shallow-focus volcanic earthquakes occurred several days before each dome-building eruption.
    2 As the number of earthquakes increases the total seismic energy is calculated. A sudden increase in seismic energy occurs a few hours before an eruption.
    3 Once an eruption is under way, shallow-focus earthquakes cease and surface events from rockfalls become dominant.
    4 Before explosive eruptions, there is an increase in shallow volcanic and harmonic tremor earthquakes.
(iii) Fig. 3.10D is a crustal section showing the positions of foci associated with (1) earthquakes following the explosive eruptions of 1980; and (2) earthquakes preceding the dome-building eruption of March 19, 1982. Account for the positions of the foci.

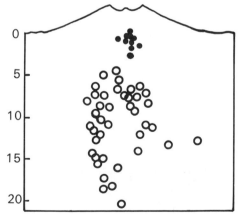

Foci of earthquakes following explosive eruptions of 1980: ○
Foci of earthquakes preceding dome-building eruption of March 19, 1982: ●

Fig. 3.10D

(e) *Gas emission studies*

Most of the gas from Mount St Helens is water vapour but emissions also include sulphur dioxide, carbon dioxide and hydrogen, with lesser amounts of helium, carbon monoxide, hydrogen sulphide and hydrogen chloride.

Account for the following observations:

    1 Rates of sulphur dioxide emission increase steadily before an explosive eruption.

    2 There is no sudden decline in sulphur dioxide emission when an explosive eruption starts.

    3 Emission rates for sulphur dioxide reached their maximum levels in mid-1980 at about 1500 tonnes per day. In 1983 they were only about 100 tonnes per day.

    4 Before a non-explosive dome-building eruption, sulphur dioxide emission increases. It remains high during the eruption then drops to pre-eruption levels when lava extrusion stops.

3.11   Identify the features shown in the photograph and say how they were formed.

SURFACE PROCESSES

4.1 Identify the feature in the photograph and describe how it was formed. The rock is dolerite.

4.2 (a) (i) What is mass movement (mass wasting)?
(ii) Which processes contribute to soil creep?
(iii) How do solifluction, earthflow, mudflow, rockslide and landslide differ?
(iv) In the triangular diagram (Fig. 4.2), identify processes W—Z.

(b) (i) Various methods can be used to measure rates of soil creep. What are the advantages and disadvantages of the following?
1 Markers such as coloured stones placed on the slope.
2 Rigid pegs inserted vertically into the slope.
3 Pliable plastic tubing buried at right angles to the slope.
4 Pegs buried at various depths with their long axes parallel to the slope contours.
(ii) Why should the buried pegs in method 4 be of the same density as the regolith?

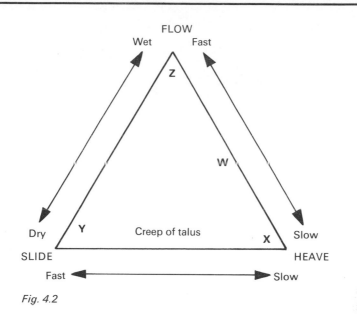

FLOW

Wet   Fast

Z

W

Dry   Y   Creep of talus   X   Slow

SLIDE   HEAVE

Fast ←————————→ Slow

Fig. 4.2

4.3 Processes contributing to denudation were studied in northern Lappland. The results are shown in Table 4.3.

Process	Average slope gradient	Vertical transport (*tonnes × metres*)	Horizontal transport (*tonnes × metres*)
Rockfall	45°	19 565	19 565
Slush and snow avalanche	30°	21 850	37 820
Debris avalanche, debris slide, mudflow	30°	96 375	166 630
Talus creep	30°	2 700	4 700
Solifluction	15°	5 300	19 800
Solution loss	30°	136 500	236 500

Table 4.3

   (a) What is solution loss?
   (b) What percentages of total vertical and horizontal transport are provided by solution loss? Why is it somewhat surprising that solution loss is the major process of denudation in this area?
   (c) What percentages of vertical and horizontal transport are provided by rapid downslope movements?
   (d) Why does rockfall have equal figures for vertical and horizontal transport?
   (e) What percentages of vertical and horizontal transport are provided by slow downslope movements?
   (f) In what climatic regions might solifluction be the dominant process of denudation?

4.4 A reservoir filled in 1900 had the following characteristics.

Capacity: $25 \times 10^4 \, m^3$
Surface area: $5.4 \times 10^4 \, m^2$
Catchment area: $3.4 \, km^2$
Mean annual precipitation: $1.65 \, m$
Annual surface runoff: $1.25 \, m$

The reservoir was drained in 1983. The sediment on the floor of the reservoir had the following properties.

Average thickness: $0.5 \, m$

Weight composition of sediment
Water: 51%
Organic matter: 10%
Inorganic sediment (*sand, silt, etc.*): 39%

Relative density (*specific gravity*) *of inorganic sediment*: 2.6
Percentage volume of inorganic sediment: 15%

(a) What is the volume of the wet sediment?
(b) What was the percentage reduction in the capacity of the reservoir?
(c) What is the volume of the inorganic sediment?
(d) What is the mass of the inorganic sediment?
(e) What is the volume of water in the sediment?
(f) What is the mass of water in the sediment?
(g) What is the mass of the organic matter in the sediment?
(h) What is the total mass of the organic and inorganic sediment?
(i) What is the percentage volume of organic matter in the wet sediment?
(j) What is the relative density (specific gravity) of the organic matter?
(k) What is the source of the organic matter?
(l) What is the average mass of inorganic sediment deposited every year?
(m) What is the average volume of inorganic sediment deposited every year?
(n) What is the average mass of inorganic sediment eroded from each km^2 of the catchment area every year? (Note that only 80% of the sediment entering the reservoir is deposited.)
(o) What is the average volume of inorganic sediment eroded from each km^2 of the catchment area every year?
(p) What is the average thickness of rock removed from the catchment area every year?
(q) What is the average mass of inorganic sediment carried by 1 m^3 of stream water entering the reservoir?
(r) Study of large rivers in the same area as the reservoir shows that 175 t of inorganic sediment are removed from each km^2 of their drainage basins every year. The river valleys were glaciated during the last Ice Age. Why does the river study suggest much higher rates of denudation than the study of the reservoir?

4.5 Fig. 4.5 shows the range of water current velocities over which sedimentary particles are picked up, transported and deposited.

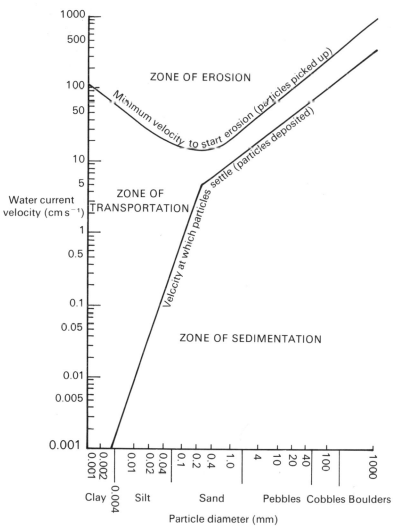

Fig. 4.5

(a) What are the ranges of current velocity within the zone of transportation for particles with the following diameters: 40, 4, 0.4, 0.04, and 0.004 mm?

(b) Which type of particle is most easily picked up by water currents?

(c) Explain why clay requires rapid currents to erode it yet it can be transported by extremely slow currents.

(d) Which type of particle is carried by the smallest range of current velocity within the zone of transportation?

(e) What is the significance of the kink in the graph showing the velocity at which particles settle?

4.6 Pebbles were collected at five sites in the valley of a large river (Fig. 4.6A). The valley was occupied by a glacier 17 000 years ago. The long (*a*), intermediate (*b*) and short (*c*) axes of each pebble were measured, and the mean axis length:

$$\left(\frac{a+b+c}{3}\right)$$

and sphericity:

$$\left(3\sqrt{\frac{bc}{a^2}}\right)$$

of each pebble were found (see Table 4.6). The roundness of each pebble was estimated by reference to a chart (Fig. 4.6B—page 28). For each group of measurements, the mean and sample standard deviation are given.

$$\text{Sample standard deviation (SD)} = \sqrt{\frac{\text{Sum of (difference between pebble value and mean value)}^2}{\text{Number in sample} - 1}}$$

$$= \sqrt{\frac{\Sigma(x-\bar{x})^2}{N-1}}$$

where: *x* is the pebble value;
$\bar{x}$ is the mean value;
N is the number in the sample;
Σ (*sigma*) means 'sum of'.

Standard deviation is a measure of how closely the measurements cluster around the mean. If the standard deviation is small, then there is little spread of results.

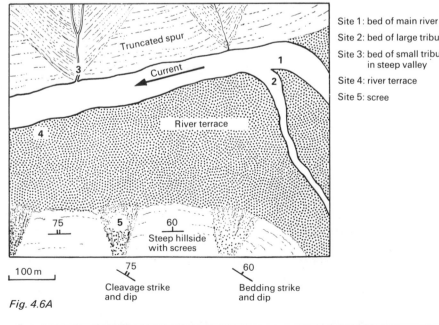

Site 1: bed of main river
Site 2: bed of large tributary
Site 3: bed of small tributary in steep valley
Site 4: river terrace
Site 5: scree

Fig. 4.6A

All measurements in cm.

Site	1	2	3	4	5
Sample size	63	58	68	65	56
Mean long axis (a)	3.42	3.89	3.96	3.52	4.89
Sample SD	0.80	0.85	0.98	1.32	1.32
Mean intermediate axis (b)	2.28	2.63	2.70	2.51	2.69
Sample SD	0.52	0.69	0.67	0.82	0.95
Mean short axis (c)	1.44	1.58	1.54	1.47	1.20
Sample SD	0.46	0.42	0.51	0.59	0.39
Mean axis length	2.38	2.70	2.74	2.50	2.93
Sample SD	0.51	0.56	0.60	0.83	0.64

Number of pebbles in each range of mean axis length

	1	2	3	4	5
1.00–1.49				3	
1.50–1.99	13	7	8	17	2
2.00–2.49	28	15	20	17	12
2.50–2.99	15	19	19	14	20
3.00–3.49	4	11	13	6	10
3.50–3.99	3	4	6	3	10
4.00–4.49		2	2	2	1
4.50–4.99				2	
5.00–5.49				1	1

	1	2	3	4	5
Mean sphericity	0.66	0.65	0.64	0.67	0.52
Sample SD	0.08	0.07	0.10	0.08	0.12

Number of pebbles in each range of sphericity

	1	2	3	4	5
0.30–0.39					5
0.40–0.49	1	1	3		22
0.50–0.59	14	13	19	13	16
0.60–0.69	29	29	22	26	8
0.70–0.79	17	13	23	21	4
0.80–0.89	2	2	1	5	1

Number of pebbles in each shape category

Roundness	Roundness index	1	2	3	4	5
Very angular	0.15			4		36
Angular	0.21			21		20
Subangular	0.30	9	8	24	4	
Subrounded	0.42	21	21	16	14	
Rounded	0.60	19	20	2	21	
Well rounded	0.85	14	9	1	26	
Mean roundness index		0.55	0.53	0.31	0.64	0.17

Table 4.6

PEBBLE ROUNDNESS CHART

Degree of roundness	Very angular	Angular	Sub-angular	Sub-rounded	Rounded	Well rounded
Appear-ance of pebbles						
Round-ness index	0.15	0.21	0.30	0.42	0.60	0.85

Fig. 4.6B

(a) How are scree pebbles formed?

(b) The scree pebbles are covered by thick growths of lichen. What does this suggest about the present rate of scree formation?

(c) How have the river terraces been formed?

(d) Why do the scree pebbles generally have smaller c axes than pebbles from other sites?

(e) Distinguish between the terms 'sphericity' and 'roundness'.

(f) How do the sphericities of a marble, a cube and a matchbox compare?

(g) How do the roundness of a marble, a cube and a matchbox compare?

(h) How are pebbles transported by a river?

(i) How does transport affect sphericity and roundness?

(j) How would the original shape of a pebble affect its final shape?

(k) Why is the sample SD of the river terrace pebble sizes much greater than that of the pebbles from the main stream?

(l) In the pebbles, what general relationships exist between:
1 mean axis length and sphericity?
and 2 mean axis length and roundness?

(m) In what way are the pebble samples unrepresentative?

4.7 (a) Fig. 4.7A shows features associated with karst landscape.

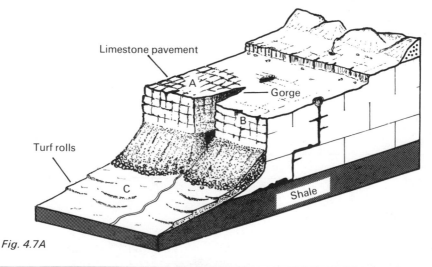

Limestone pavement

Gorge

Turf rolls

Shale

A

B

C

Fig. 4.7A

(i) Briefly describe the processes taking place at A, B and C.
(ii) In the scree, why do the particles show a general increase in size from the top towards the bottom of the slope?
(iii) How are gorges formed in limestone areas?

(b) Fig. 4.7B shows the monthly levels of dissolved calcium carbonate in water from a spring in a limestone area.

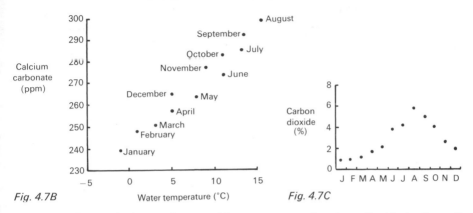

Fig. 4.7B

Fig. 4.7C

Fig. 4.7C shows the monthly percentage of carbon dioxide in the soil atmosphere above the limestone.

(i) How can you tell from Fig. 4.7B that calcium carbonate solubility does not depend entirely on water temperature?
(ii) What factor in addition to water temperature causes a change in calcium carbonate solubility?
(iii) Why does the carbon dioxide level in the soil atmosphere vary in the way shown in Fig. 4.7C?
(iv) Explain why limestone is more soluble in rain water than in pure water.

4.8 (a) Identify the coastal features in the middle of the photograph. Why do these features have a step-like form?

(b) In East Fife, raised shorelines slope as shown in Fig. 4.8B.
  - (i) Why are the shorelines tilted?
  - (ii) Why do the younger shorelines generally extend further west than the older shorelines?
  - (iii) Shoreline 6 has a gradient about half that of shoreline 1. What does this indicate about the amounts of early and late isostatic movements?
  - (iv) If the shorelines were extrapolated eastwards, what would happen to their order?
    What is the relationship between the rates of upward movement of the land and sea in the west and in the east?
  - (v) When shoreline 1 was being formed, it was thought to be 180 km from the main centre of ice accumulation in the Grampian Highlands. If this shoreline has a constant gradient, how much isostatic uplift have the Grampian Highlands undergone?

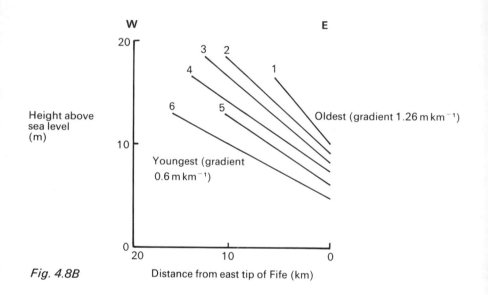

Fig. 4.8B

4.9 Pebbles may be classified as spheres, discs, rods and blades. Where a is the long axis, b the intermediate axis and c the short axis, the shapes have the following axial ratios:

sphere	$b/a > 0.67$	$c/b > 0.67$
disc	$b/a > 0.67$	$c/b < 0.67$
rod	$b/a < 0.67$	$c/b > 0.67$
blade	$b/a < 0.67$	$c/b < 0.67$

Samples of pebbles from three positions on a beach were measured to determine their shapes. The results are shown in the histograms in Fig. 4.9.

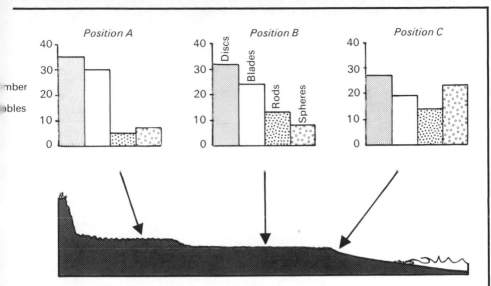

Fig. 4.9

(a) Account for the different shape distributions at positions A, B and C.
(b) The pebbles were of four types:
dolerite, 117 pebbles; sandstone, 53 pebbles; gneiss, 34 pebbles; and slate, 36 pebbles.
Suggest possible relationships between the rock types and the pebble shape distributions at A, B and C.

4.10 (a) (i) Identify features A—K in the diagram of a glaciated landscape (Fig. 4.10A). How were features A—D formed?

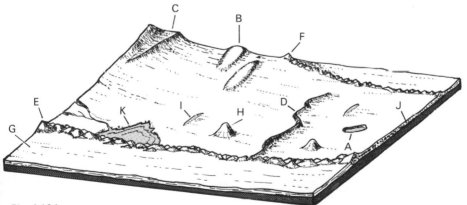

Fig. 4.10A

(ii) How could you distinguish between the sediments of the moraines and esker?
(iii) How are varved sediments formed? What is their stratigraphical value?

(b) Fig. 4.10B shows features in glaciofluvial sediments.

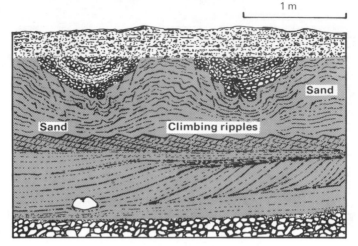

Fig. 4.10B

(i) Which processes may have contorted the bedding of the upper sand?
(ii) How have the climbing ripples formed?
(iii) How has the boulder been deposited within the lower sand?
(c) Fig. 4.10C shows a soil surface pattern of stone circles and stripes.

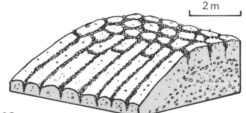

Fig. 4.10C

(i) Under what climatic conditions do these features form?
(ii) How have the circles and stripes formed?
(d) Fig. 4.10D shows change in the heights of corries (cirques) in part of the Scottish Highlands. The heights of the mountains do not change across the area.

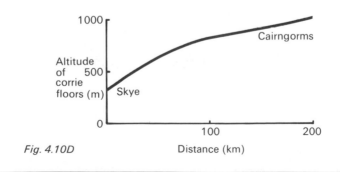

Fig. 4.10D

(i) State one factor which could affect the altitude at which the corries form. How does this factor affect the altitude of corrie formation?

(ii) With the aid of diagrams, explain how corries form.

4.11 (a) (i) Identify features A—E in the diagram of an arid landscape (Fig. 4.11A).

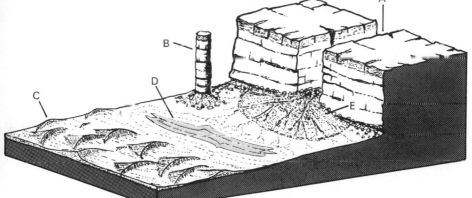

Fig. 4.11A

(ii) How are barchan and longitudinal (seif) dunes formed?

(iii) Why do valley sides tend to be steeper in arid regions than in humid regions?

(iv) What types of deposit are normally found in playas?

(b) In arid regions, salt weathering is a common cause of rock disintegration. Every day, sandstone cubes were soaked in saturated salt solutions and then dried. The weight changes produced by various salts are shown in Fig. 4.11B.

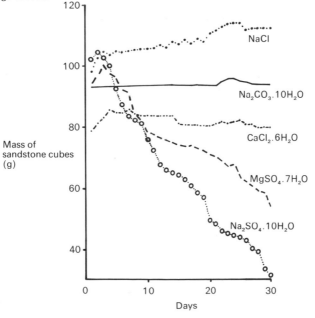

Fig. 4.11B

(i) Which processes contribute to salt weathering?

(ii) Why are some salts more effective than others at causing rock disintegration?

(iii) Why do the sandstone cubes generally show initial increases in weight?

(c) Fig. 4.11C shows weight changes in equal-sized cubes of various rocks which, every day, were soaked in saturated sodium sulphate solution and dried.

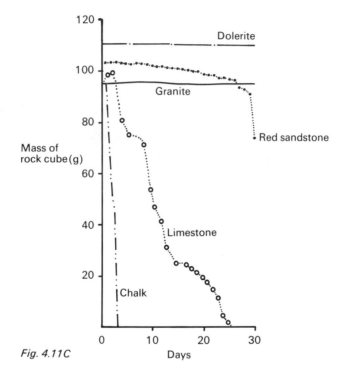

Fig. 4.11C

(i) Why was sodium sulphate chosen for this experiment?

(ii) Why does chalk break down much more quickly than granite?

(iii) Why does the red sandstone show a sudden weight loss after about 29 days?

SEDIMENTARY ROCKS

5.1 Complete the chart, Fig. 5.1, to show a classification of sedimentary rocks.

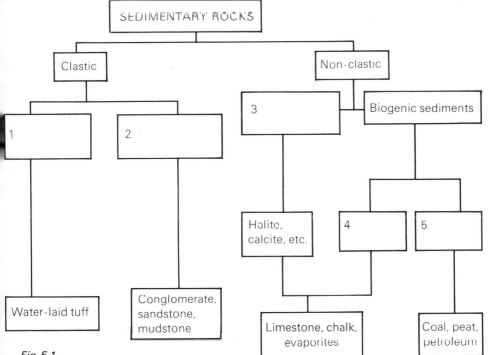

Fig. 5.1

5.2 Table 5.2 shows chemical analyses of sandstone, shale, granite and basalt. (Oxides in weight per cent.)

	Sandstone	Shale	Granite	Basalt
SiO_2	95.4	55.1	70.8	49.0
Al_2O_3	1.1	16.3	14.6	18.2
Fe_2O_3	0.4	4.2	1.6	3.2
FeO	0.2	1.9	1.8	6.0
MgO	0.1	2.5	0.9	7.6
CaO	1.6	4.7	2.0	11.2
Na_2O	0.1	0.7	3.5	2.6
K_2O	0.2	3.0	4.2	0.9
H_2O	0.3	5.2	0.3	0.2
CO_2	1.1	4.0	—	—

Table 5.2

Account for the following observations.
(a) The sandstone contains more SiO_2 than the shale.
(b) Shale contains much more Al_2O_3 than sandstone.
(c) In the igneous rocks, FeO exceeds Fe_2O_3. In the sedimentary rocks, Fe_2O_3 exceeds FeO. Shale contains more Fe_2O_3 than the other rocks.
(d) Total MgO in the igneous rocks exceeds total MgO in the sedimentary rocks. Total CaO in the igneous rocks exceeds total CaO in the sedimentary rocks.
(e) The sedimentary rocks contain much less Na_2O than the igneous rocks.
(f) The shale contains much more K_2O than Na_2O.
(g) The shale contains much more H_2O than the other rocks.
(h) The sedimentary rocks contain CO_2. The igneous rocks have no CO_2.

5.3 (a) In the rock cycle, Fig. 5.3, which words or phrases from the following list are represented by letters A—K?
Denudation; deposition; diagenesis; heat and pressure; igneous rocks; melting; metamorphic rocks; rocks moved to surface; sedimentary rocks; solidification; transport.

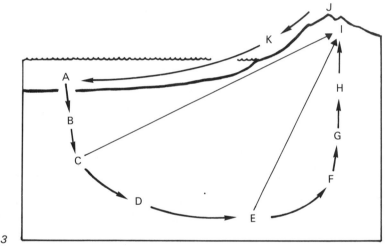

Fig. 5.3

(b) Explain the meanings of *diagenesis*, *lithification* and *authigenic minerals*.
(c) How does connate water differ from juvenile water?

5.4 Samples of sand from different environments were passed through sets of sieves. The results are shown in Table 5.4A.

The sand grains had the following properties.
River sand: angular to subrounded; about half of the grains have polished surfaces.
Desert sand: well-rounded; frosted.
Beach sand: subrounded to well-rounded; all grains are polished.
Sand X: very angular to subrounded; some of the subrounded grains show signs of polish.

Sieve mesh size (μm)	Phi (φ) value	River sand	Desert sand	Beach sand	Sand X (source unknown)
		Weight per cent of sand retained on each sieve			
8000	−3				4
4000	−2	7			4
2000	−1	12			16
1000	0	21	4		41
500	1	32	76		28
250	2	18	16	23	4
125	3	6	2	71	2
63	4	4	2	6	1

Table 5.4A

(a) Plot cumulative frequency percentage curves for the four sands using the method shown in the example.

Example:

Sieve mesh size (μm)	Weight % of sand on each sieve	Cumulative weight % of sand above each sieve
2000	0	0
1000	5	5
500	12	17
250	31	48
125	23	71
63	18	89
32	11	100

Table 5.4B

Cumulative frequency curve with φ values shown in Fig. 5.4A.

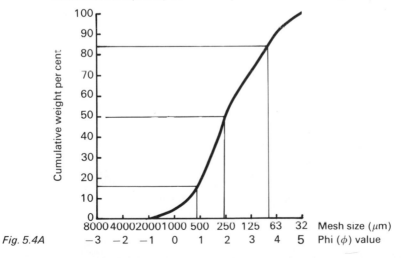

Fig. 5.4A

Phi values are related to mesh sizes by the following equation:

$$\text{mesh size} = 1000 \times 2^{-\phi}$$

(b) (i) How does a well-sorted sediment differ from a poorly-sorted sediment?

(ii) The coefficient of sorting can be found from the cumulative frequency curve using the relationship:

$$\text{sorting} = \frac{\phi_{84} - \phi_{16}}{2}$$

where ϕ_{84} and ϕ_{16} are the phi values corresponding to the 84th and 16th percentiles (see Fig. 5.4A). The 84th and 16th percentiles are chosen because they are one standard deviation away from the median value (ϕ_{50}). Sorting is a measure of the standard deviation of the grain size distribution.

Find the sorting coefficients of the four sediments.

(c) Skewness is a measure of the asymmetry of the grain size distribution (Fig. 5.4B).

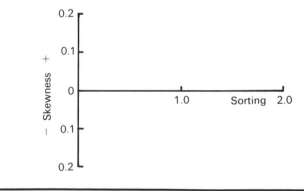

Fig. 5.4B   Unskewed   Negatively skewed   Positively skewed

A normal distribution is not skewed; a distribution in which coarser-grained particles predominate is negatively skewed; and a distribution in which finer-grained particles predominate is positively skewed.

(i) Why do sediments in the upper course of a river often show negative skewness while those in the lower course often show positive skewness?

(ii) Values of skewness can be found from the cumulative frequency curves using the relationship:

$$\text{skewness} = \frac{(\phi_{84} + \phi_{16}) - 2\phi_{50}}{2}$$

Find the values of skewness for the four sediments.

(d) On a graph like that shown in Fig. 5.4C, plot skewness against sorting for the sediments.

Fig. 5.4C

(e) Considering grain properties and values of sorting and skewness, compare Sand X with the other sediments and discuss a possible environment of deposition. Why is your interpretation necessarily tentative? What other evidence would you need to allow you to come to firmer conclusions about the environment of deposition of Sand X?

5.5 Identify the features in the photograph and say how they were formed.

5.6 Identify the features shown in the photograph and say how they were formed. The rock is greywacke.

5.7 Fig. 5.7 shows rocks deposited in a sedimentary basin.

(a) Give an account of the sedimentary history of the basin.
(b) Explain why the fossil coral reefs are found only at the edge of the basin.

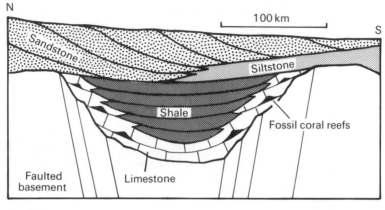

Fig. 5.7

5.8 The map (Fig. 5.8) shows the geography of an area which has a hot, dry climate. Heavy rain sometimes falls in the mountains but no rivers reach the sea. The wind blows persistently from the land towards the sea. An arm of the sea forms a partly enclosed basin. The inflow of water to the basin exceeds the outflow.

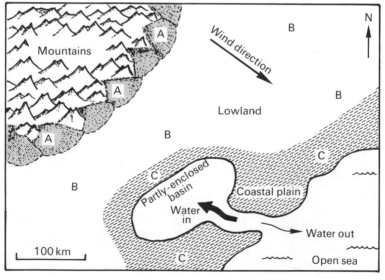

Fig. 5.8

(a) What types of sediment would probably be deposited in the following areas?
Close to the mountains (area A); on the lowlands (area B); on the coastal plain (area C); in the open sea.

(b) What type of sediment would be deposited in the partly enclosed basin?

(c) Sea water contains the following salts.

Salt	Percentage of dissolved solids
$NaCl$	78.04
$MgCl_2$	9.21
$MgSO_4$	6.53
$CaSO_4$	3.48
KCl	2.11
$CaCO_3$	0.33
$MgBr_2$	0.25

Table 5.8A

The most common evaporite sequence consists of the following salts.

Top	$NaCl$	20%
↑	$CaSO_4$	60%
Bottom	$CaCO_3$	20%

Table 5.8B

Explain why common evaporite sequences do not necessarily contain a complete sequence of sea salts.

5.9 Fig. 5.9 shows sediments deposited in nearshore environments.

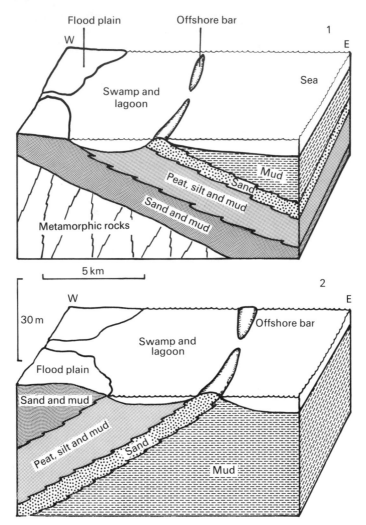

Fig. 5.9

(a) How do the sedimentary sequences differ? How have these differences been produced?
(b) By reference to Fig. 5.9, explain the meanings of 'diachronism' and 'sedimentary facies'.
(c) In which sediments in Fig. 5.9 would you expect to find cross bedding and mudcracks?

EARTH PHYSICS

6.1 (a) On what rock properties do the velocities of P- and S-waves depend? Why do S-waves not pass through liquids?

(b) Fig. 6.1 shows how P- and S-wave velocities change with depth. Account for the following observations.

1 Wave velocity increases gradually from the surface to the base of the crust.

2 There is a sudden increase in wave velocity at the Moho.

3 Wave velocity decreases in the area between depths of about 50 and 250 km.

4 Gradual increases in velocity occur at depths of about 400, 700 and 1050 km.

5 At the core-mantle boundary, S-wave velocity falls to zero. P-wave velocity falls from 13.6 to 8.1 km s^{-1}.

6 P-wave velocity increases sharply at a depth of about 5150 km.

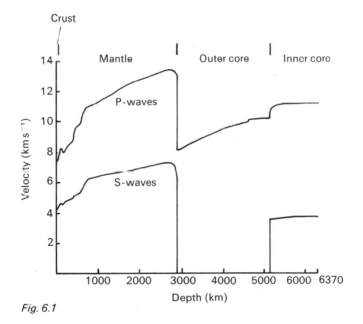

Fig. 6.1

6.2 (a) Fig. 6.2A shows a recording of P- and S-waves from a near-surface earthquake received at four seismometer stations A—D. The P-waves travel at $5.8 \, km \, s^{-1}$.

(i) How far is each recording station from the earthquake focus?

(ii) What is the velocity of the S-waves?

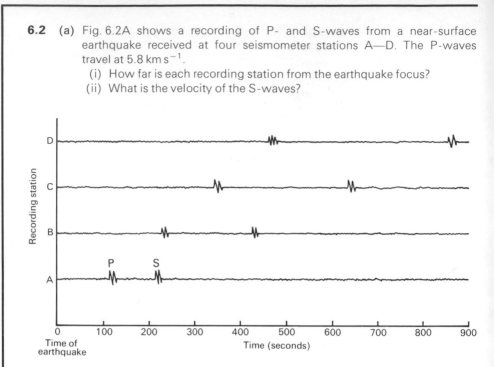

Fig. 6.2A

(iii) The map (Fig. 6.2B) shows the positions of the recording stations. On a tracing of the map, mark an 'X' at the position of the earthquake epicentre.

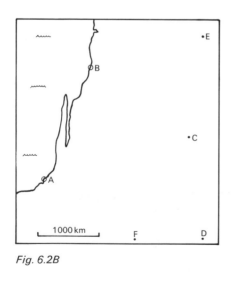

Fig. 6.2B

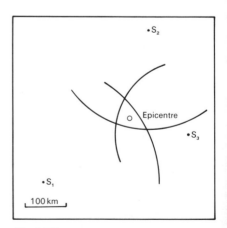

Fig. 6.2C

(iv) How long would it take for P-waves to reach recording stations E and F?

(v) For stations A—D, draw a graph of time interval between the arrival of the P- and S-waves against distance from the earthquake focus. From your graph find the time lag between the arrival of P- and S-waves for every 1000 km from the focus.

(b) The S—P time interval can be used to find the distance between the focus and the recording station. Fig. 6.2C shows circles, centred on recording stations, of radii equal to the distances between the focus and the seismometers. The earthquake focus lies at a depth of 100 km. Why do the circles not intersect at the epicentre?

(c) The depth of focus of an earthquake can be estimated from the differences in intensity (measured in terms of acceleration from seismograms) at the epicentre and at a point on an isoseismal line. Intensity is proportional to $1/r^2$, where r is the distance from the focus. The intensity at E (the epicentre) is proportional to $1/h^2$, where h is the focal depth. Similarly, the intensity at S (the seismometer) is proportional to $1/r^2$. From this we can say:

$$\frac{\text{intensity at S}}{\text{intensity at E}} = \frac{h^2}{r^2} = \sin^2 \theta$$

From this equation we can find θ (the angle between r and the ground surface) and:

$$h = d \tan \theta$$

where d is the epicentral distance.

Find the focal depths of earthquakes with the characteristics given in Table 6.2A.

	Epicentral distance (km)	Ground acceleration (ms^{-2}) at epicentre	Ground acceleration (ms^{-2}) at seismometer
(i)	300	5.0	0.25
(ii)	350	7.5	5.0
(iii)	400	7.5	0.1
(iv)	700	5.0	0.1

Table 6.2A

(d) When an earthquake wave crosses a discontinuity, it is refracted and its velocity changes (Fig. 6.2D overleaf). The angular and velocity relationships are given by the equation:

$$\frac{\text{velocity of incident wave } (V_I)}{\text{velocity of refracted wave } (V_R)} = \frac{\sin \theta_I}{\sin \theta_R}$$

where θ_I is the angle of incidence and θ_R is the angle of refraction.

(i) Complete Table 6.2B overleaf to show wave velocities and angles of incidence and refraction.

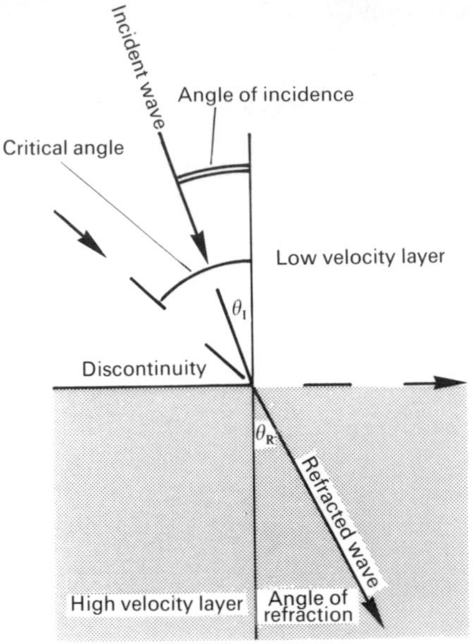

Fig. 6.2D

	Velocity of incident wave $(km\,s^{-1})$	Velocity of refracted wave $(km\,s^{-1})$	Angle of incidence	Angle of refraction
P-wave entering mantle from crust	7.0	8.0		90°
P-wave leaving mantle to enter crust	8.0	7.0	45°	
P-wave entering core from mantle	13.6		90°	36.5°
P-wave leaving core to enter mantle		13.6	23°	

Table 6.2B

(ii) Complete the sentences by inserting the missing words.
When earthquake waves pass from a medium in which they travel _____ into one in which they travel _____, they are refracted away from the normal. When waves pass from a medium in which they travel _____ into one in which they travel _____, they are refracted towards the normal.

(iii) What is the *critical angle of incidence*?

(iv) What is the *angle of reflection*?

(v) What will happen to a wave from the crust which strikes the Moho at an angle of incidence greater than the critical angle?

6.3 Seismometers can be used to monitor distant underground nuclear explosions. Fig. 6.3 shows the pattern of P- and S-waves given off by explosions and earthquakes. In addition, both produce surface waves.

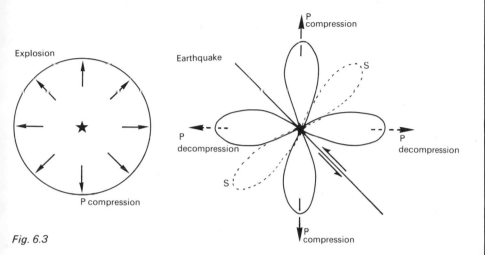

Fig. 6.3

(a) From Fig. 6.3, how do explosions differ from earthquakes? What other differences would a seismologist use to distinguish between them?
(b) How do body waves differ from surface waves?
(c) Table 6.3 gives values of magnitude measured from the body and surface waves of earthquakes and nuclear explosions.

Earthquakes		Nuclear explosions	
Surface wave magnitude	Body wave magnitude	Surface wave magnitude	Body wave magnitude
6.50	5.90	4.25	5.50
5.25	5.00	4.30	5.60
4.50	4.20	4.30	5.70
4.90	5.00	4.00	5.60
4.75	4.75	4.10	5.60
5.00	4.50	4.25	6.00
4.00	4.00	4.50	6.10
5.25	5.25	5.00	6.40
5.60	5.50	5.20	6.75
5.20	5.50	4.75	6.25
6.00	6.00	4.50	5.75
5.70	4.70	4.80	6.50
6.25	5.60		
5.75	5.25		
5.00	4.50		

Table 6.3

Plot body wave magnitudes against surface wave magnitudes for the earthquakes and nuclear explosions. What differences between earthquakes and nuclear explosions are shown by your graph?

(d) A nuclear device may be detonated in a rigid rock such as granite, or in soft material such as alluvium. How would such a difference in surrounding material affect energy release from the explosion?

6.4 Meteorites are of three main types:

1 Irons: Consist of Fe–Ni alloy with 4–20% Ni. Solidified from a melt at 1400°C. Cooled at a rate of 1–10°C per million years. Make up 5.7% of falls.

2 Stony irons: Consist of roughly equal amounts of Fe–Ni alloy and silicates such as olivine, pyroxene and plagioclase. 1.5% of falls.

3 Stones: 92.8% of falls. Two types:

  (a) Achondrites: Consist of olivine, pyroxene and plagioclase. Most look like peridotite but some resemble basalts. They have crystallized from silicate melts.

  (b) Chondrites: Make up over 85% of stones. Consist of olivine, pyroxene and plagioclase with some Fe–Ni alloy and FeS. Contain rounded silicate chondrules which represent condensed droplets of original nebular material. Chondrites have never been molten but most show signs of metamorphism. Carbonaceous chondrites consist of hydrated minerals, olivine, Fe–Ni alloy and organic matter. They are the least altered of all meteorites.

(a) From which part of the solar system do meteorites come?

(b) Why are iron meteorites the most common finds even though they provide only 5.7% of falls?

(c) Meteorites are thought to have been formed by the break-up of miniplanets or planetesimals. What evidence suggests that volcanic activity took place on planetesimals?

(d) What evidence suggests that iron meteorites formed at depths of about 200 km within planetesimals?

(e) What evidence suggests that the planetesimals had differentiated (formed layers of differing composition and density) before fragmentation?

(f) From which parts of layered planetesimals would the different types of meteorite have come?

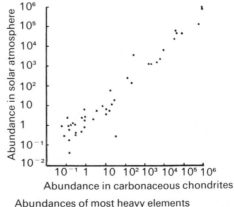

Fig. 6.4   Abundances of most heavy elements compared with silicon which is set at 10^6 atoms.

(g) The graph (Fig. 6.4) compares the abundance of elements in carbonaceous chondrites with that in the solar atmosphere. What general relationship is shown by the graph?

(h) In what ways does study of meteorites help improve our knowledge of the Earth?

6.5 (a) The gravitational force of attraction between two objects is proportional to the product of their masses. It would be expected, therefore, that a plumb-bob would be pulled towards a mountain range (Fig. 6.5A). However, mountain ranges exert less gravitational force than expected. This suggests that there is a mass deficiency associated with the mountains.
Suggest two ways in which such a mass deficiency might occur.

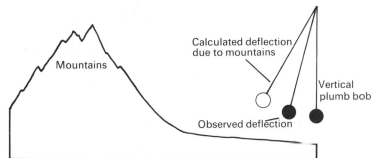

Fig. 6.5A

(b) Why are oceanic ridges higher than ocean basins?

(c) In Fig. 6.5B, an oceanic volcano is erupting magma of density 2.7 g cm^{-3}. The magma comes from a depth of 50 km. What height above the sea floor would the highest cone be which would remain in isostatic equilibrium?

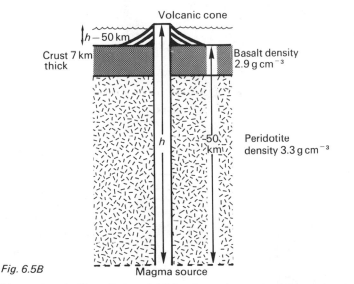

Fig. 6.5B

(d) Mauna Loa in Hawaii rises 9.75 km from the sea bed. Assuming it to be in isostatic equilibrium, from what depth does its magma come?

(e) What kind of gravity anomaly would develop if an excess of volcanic material came onto the sea floor? What would happen to the volcano as a result?

(f) What are seamounts, guyots and atolls and how are they formed?

(g) Many volcanic islands sink as isostatic equilibrium is re-established. What other process may carry volcanic islands below the sea?

6.6 (a) On separate sheets of tracing paper, draw polar wandering curves for continents A and B (Fig. 6.6).

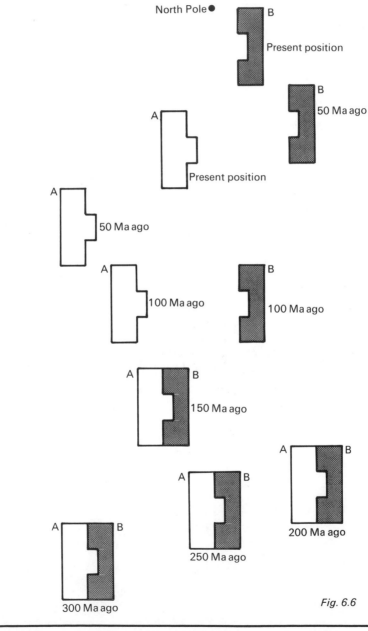

Fig. 6.6

(b) Superimpose your tracings. For what period of time do the polar wandering curves overlap? When do the curves diverge?
(c) Why are polar wandering curves badly-named?
(d) In what way does study of polar wandering curves support the theory of continental drift?

6.7 Table 6.7 shows readings of gravity and magnetic anomalies taken during a marine traverse on the continental shelf north of the Shetland Islands.

Distance along traverse (km)	0	5	10	15	20	25	30	35	40	45	50
Gravity anomaly (mgal)	0	3.5	9	21	41	53	40	18	11	5	1
Magnetic anomaly (nanotesla, nT) All values negative.	35	90	160	230	320	600	510	420	405	320	195

Table 6.7

(a) Draw graphs of gravity anomaly against distance and magnetic anomaly against distance.
(b) What is the cause of the positive gravity anomaly?
(c) The size of the gravity anomaly suggests a density contrast with the country rock of $0.3\,g\,cm^{-3}$. What is the probable density of the body producing the anomaly?
(d) What is the cause of the high negative magnetic anomaly?
(e) What type of body could cause the observed anomalies?
(f) What is the significance of the extremely high negative magnetic anomaly in the centre of the traverse?

6.8 **(a)** (i) How does heat from inside the Earth reach the surface?
(ii) The amounts of heat generated by crust and mantle rocks are as follows.

Average continental crust $\quad 1.5–3.0\,\mu J\,m^{-3}\,s^{-1}$
Average oceanic crust $\quad\quad\ \ 0.4\,\mu J\,m^{-3}\,s^{-1}$
Average upper mantle $\quad\quad\ 0.02\,\mu J\,m^{-3}\,s^{-1}$

Why do these rock types have very different heat-producing capacities?
(iii) In addition to heat produced by long-lived radioisotopes, what other sources may contribute to the Earth's heat?
(b) Table 6.8 overleaf shows average values of heat flow from continental and oceanic regions.

	Continental crust	Oceanic crust
Heat flow through Earth's surface ($J\,m^{-2}\,s^{-1}$)	53×10^{-3}	62×10^{-3}
Heat flow through Moho ($J\,m^{-2}\,s^{-1}$)	28×10^{-3}	57×10^{-3}

Table 6.8

(i) What percentage of heat escaping at the surface of continents comes from the mantle? What percentage of heat escaping from the ocean floor comes from the mantle?

(ii) Suggest why the mantle under the oceans contributes more to surface heat flow than the mantle under the continents.

(iii) The graph (Fig. 6.8) shows how heat flow changes as oceanic and continental crusts become older.

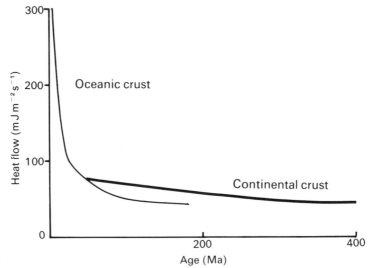

Fig. 6.8

Why does heat flow fall rapidly from oceanic crust but only slowly from continental crust?

METAMORPHIC ROCKS

7.1 Identify the following metamorphic rocks from their descriptions.

(a) Irregularly banded; dark with pale veins and lenses. Looks like a mixture of granitic material and dark gneiss. Shows signs of having been partly molten. Main minerals: feldspar, quartz, hornblende, garnet, mica.

(b) Fine-grained, metamorphosed pelitic sediment. Splits easily along cleavage planes. Bedding may be seen on cleavage planes. Main minerals: quartz, muscovite, chlorite, feldspar.

(c) Fine- or medium-grained. Breaks unevenly into splintery fragments. No foliation. Main minerals: feldspar, quartz, andalusite, cordierite, biotite, hornblende, pyroxene.

(d) Shiny grey-green. Fine- or medium-grained. Well developed foliation. Foliation planes often have a wrinkled appearance. Metamorphosed pelitic sedimentary rock. Main minerals: quartz, green chlorite, muscovite, feldspar.

(e) Green, grey or black. Medium- or coarse-grained. Irregularly veined. Metamorphosed ultrabasic igneous rock. Main minerals: serpentine, chlorite, talc, calcite, iron ore.

(f) Coarse-grained. Grains may be of roughly the same size. Unfoliated but may show large-scale banding. No hydrated minerals. Formed at high temperature. Main minerals: pyroxene, plagioclase, quartz, garnet.

7.2 Briefly describe the processes which produced the metamorphic textures shown in Fig. 7.2 (a)–(f).

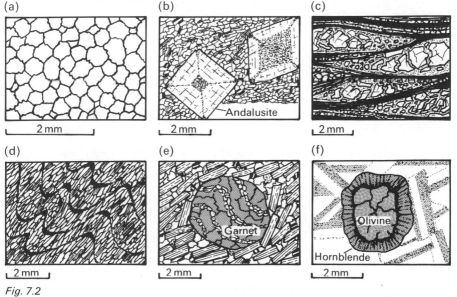

Fig. 7.2

7.3 Fig. 7.3 shows thin sections made from rocks S1, S2 and R in the photograph.
  (a) (i) Suggest a name for each rock. Give reasons for your choices of name.
      (ii) Why do rocks S1 and S2 have different textures?
  (b) Outline the sequence of events which produced the rocks and relationships shown in the photograph.

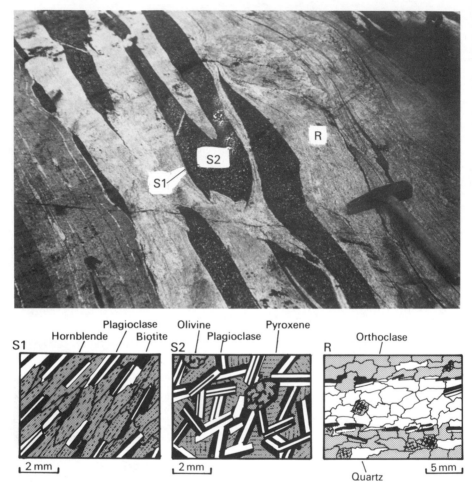

Fig. 7.3

7.4 The map (Fig. 7.4) shows the route taken during a geological traverse in south-west England.

  (a) (i) Each of the following rocks was collected at one of positions 1–7. At what position was each rock collected?
      Calcium silicate hornfels; chiastolite shale; granite; hornblende plagioclase hornfels; shale; spotted shale; tourmaline granite.
      Briefly state how spotted shale, chiastolite shale, calcium silicate hornfels and hornblende plagioclase hornfels have formed.
      (ii) Explain the meanings of: stream-tin; tor; sheet joints.
  (b) Give a very brief outline of the geological history of the area.

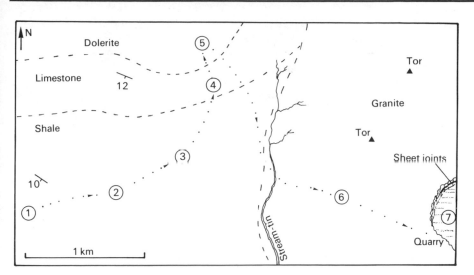

Fig. 7.4

7.5 (a) Explain the meanings of: metamorphic grade; index mineral; metamorphic zones; isograd.

(b) (i) The map (Fig. 7.5) shows an area of metamorphic rocks. On a tracing of the map, draw in the isograds for chlorite, biotite, garnet, staurolite, kyanite and sillimanite.

(ii) In which direction does the metamorphism increase in intensity?

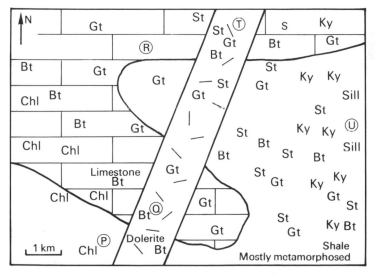

Index to minerals: Chl: chlorite Bt: biotite Gt: garnet St: staurolite
Ky: kyanite Sill: sillimanite

Fig. 7.5

(c) Table 7.5 shows mineral assemblages in metamorphosed shale, limestone and dolerite found within different zones.

Zone	Metamorphosed shale	Metamorphosed limestone	Metamorphosed dolerite
Chlorite	Chlorite	Chlorite	Chlorite
	Quartz	Calcite	Plagioclase
	Plagioclase	Plagioclase	
	Mica		
Biotite	Biotite	Chlorite	Biotite
	Quartz	Calcite	Chlorite
	Plagioclase	Plagioclase	Plagioclase
Garnet	Garnet	Garnet	Garnet
	Biotite	Calcite	Biotite
	Quartz	Epidote	Plagioclase
	Plagioclase	Hornblende	
Staurolite	Staurolite	Garnet	Garnet
	Garnet	Epidote	Hornblende
	Mica	Plagioclase	Plagioclase
	Quartz	Hornblende	
	Plagioclase		
Kyanite	Kyanite	Garnet	Hornblende
	Garnet	Hornblende	Plagioclase
	Mica	Plagioclase	
	Quartz		
	Plagioclase		
Sillimanite	Sillimanite	Garnet	Hornblende
	Garnet	Pyroxene	Plagioclase
	Mica	Plagioclase	
	Quartz		
	Plagioclase		

Table 7.5

What mineral assemblages would you expect to find at positions P–U (Fig. 7.5)? Why do mineral assemblages differ within a zone?

7.6 (a) Fig. 7.6 shows the pressure and temperature stability fields of andalusite, kyanite and sillimanite.

(i) Under a normal geothermal gradient of $0.03\,°C\,m^{-1}$, at what depth would a temperature of $120\,°C$ be attained? Which phase would crystallize at this temperature and pressure?

(ii) Under a normal geothermal gradient, at what depth would a temperature of $360\,°C$ be attained? Which phase would crystallize at this temperature and depth?

(b) The Dalradian rocks of the Scottish Highlands show regional metamorphism of two types:

1 *Barrow-type metamorphism* where the highest grades are represented by the kyanite and sillimanite zones.

2 *Buchan-type metamorphism* where the highest grades are represented by the andalusite and sillimanite zones.

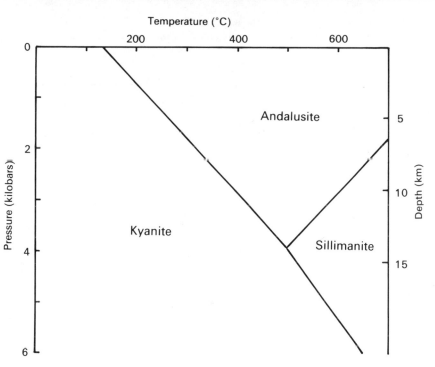

Fig. 7.6

(i) How do pressure conditions in the two types of metamorphism differ?

(ii) Sillimanite-grade Barrow-type metamorphism took place at a depth of 12 km and at a temperature of about 600 °C. What was the geothermal gradient? How many times greater than the present normal gradient of $0.03 °C\,m^{-1}$ was the geothermal gradient during Barrow-type metamorphism?

(iii) Sillimanite-grade Buchan-type metamorphism took place at a depth of 4 km and at a temperature of about 800 °C. What was the geothermal gradient? How many times greater than the present normal gradient was this geothermal gradient?

(iv) How could the relatively high geothermal gradient associated with Barrow-type metamorphism be produced?

(v) How could the very high geothermal gradient associated with Buchan-type metamorphism be produced?

7.7 **(a)** What is a metamorphic facies?

(b) Fig. 7.7 shows the pressure and temperature fields of various facies.

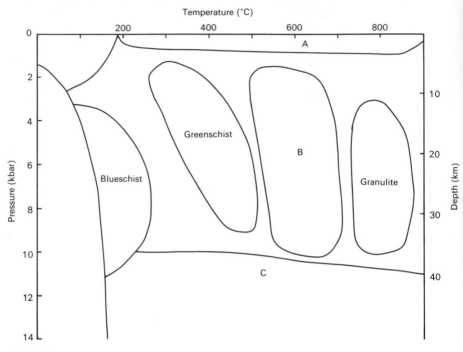

Fig. 7.7

(i) What are the approximate ranges of pressure and temperature within which granulite and blueschist facies metamorphism take place?

(ii) Which facies are represented by areas A, B and C?

(iii) Select the three correct statements from the following list:

A Granulite facies metamorphism is typical of constructive plate margins.

B Blueschist (glaucophane schist) facies metamorphism may occur in subduction zones.

C Amphibolite facies metamorphism cannot cause partial melting of rocks.

D Greenschists are produced by high grade regional metamorphism.

E With increased pressure, blueschists may be converted to eclogites.

F Granulite facies metamorphism may cause partial melting of rocks.

G Eclogite facies metamorphism can take place only in the mantle.

H The facies principle refers only to the metamorphism of sedimentary rocks.

(c) The facies idea makes no allowance for the presence of fluids other than water and it does not allow for variations in fluid pressures. How may assumptions such as these affect the definition of facies pressure and temperature fields?

THE MOVING EARTH

8.1 Pebbles from undeformed and deformed conglomerate were measured. The ratios of the pebble axes are shown in Table 8.1. In the deformed conglomerate, the short axes of the pebbles are perpendicular to the bedding. Sample A came from an undeformed conglomerate; sample B came from the lower, non-inverted limb of a recumbent fold; and sample C came from the upper, inverted limb of the same fold.

	Ratio of pebble axes		
	Long axis (*a*) :	Intermediate axis (*b*) :	Short axis (*c*)
Sample A *161 pebbles*	1.3 :	1.1 :	1.0
Sample B *80 quartzite pebbles*	33.0 .	25.0 .	1.0
31 quartz pebbles	18.5 :	14.0 :	1.0
Sample C *24 quartzite pebbles*	51.0 :	44.5 :	1.0
8 quartz pebbles	19.0 :	16.0 :	1.0

Table 8.1

When an object is deformed the strain is given by:

$$\text{strain} = \frac{\text{length after strain} - \text{length before strain}}{\text{length before strain}}$$

$$= \frac{\text{length after strain}}{\text{length before strain}} - 1$$

(a) Assuming that the long, intermediate and short pebble axes before deformation become the same axes after deformation, calculate the strains parallel to the axes of the pebbles in samples B and C. Pebble volumes do not change during deformation.

(b) If the beds undergo the same strain as the quartzite pebbles, what are the changes in bedding thickness in both limbs of the fold?

(c) Why is it not necessarily correct to assume that the beds and the quartzite pebbles have been equally strained?

8.2 Identify the features in the photograph and say how they were formed. The rock is basalt.

8.3 In part of a coalfield, the ground surface is level and the beds are horizontal and unfolded. Fossils found in boreholes P–S and Coal Measure zones based on non-marine bivalves are shown in the Tables 8.3A and 8.3B. The boreholes lie 200 m apart on a straight line.

| Depth below surface (m) | Fossils found in boreholes | | | |
	P	Q	R	S
50			Fault breccia	
100	*Antraconaia pulchra*			
150		*Anthraconaia prolifera*		
200		*Anthraconauta tenuis*	*Anthraconauta tenuis*	
300	*Anthraconaia lenisculata*		*Anthraconaia pulchra*	*Anthraconauta tenuis*
450	Fault breccia			
500	*Anthraconaia lenisculata*	*Anthraconaia lenisculata*	*Anthraconaia lenisculata*	

Table 8.3A

60

Coal Measure zone fossils

Part of Coal Measures	Zone
Upper	*Anthraconaia prolifera* *Anthraconauta tenuis* *Anthraconauta phillipsi*
Middle	*Anthraconaia pulchra* *Anthracosia similis* —*Anthraconaia modiolaris*—
Lower	*Carbonicola communis* *Anthraconaia lenisculata*

Table 8.3B

Draw a geological section to show zones and faults.
Label: 1 The fault with the greatest vertical displacement;
      2 a reverse fault.

8.4 Fig. 8.4A shows the positions of four vertical rock faces beside a horizontal road. Exposures A–D are illustrated overleaf.

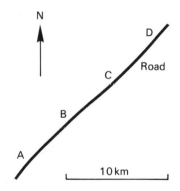

Fig. 8.4A

Exposure A

Sand and gravel
2 m thick

Limestones and shales
with ammonites
Dip 10° to NW
4 m thick

Limestones and shales with
Lithostrotion and *Dictyoclostus*
[*Productus*]
Cross-bedded sandstone
Conglomerate

Greywackes and shales
with *Monograptus*

Cleavage strikes NW-SE

Exposure B

Peat

Shales and limestones with
ammonites and belemnites

8 m thick

Dip 10° to NW

Limestone with
crinoids and trilobites

Shale
Sandstone   Dip 10° to SE
Conglomerate

Greywackes and shales
No fossils
Dip 70° to NE

Medium-grained black
igneous rock

Exposure C

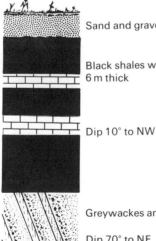

Sand and gravel 1 m thick

Black shales with limestones
6 m thick

Dip 10° to NW

Greywackes and shales

Dip 70° to NE

Exposure D

Sand with cross bedding
and ripple marks

3 m thick

Shales and limestone

4 m thick

Dip 8° to NW

Siltstone

Conglomerate

Pegmatite (quartz and feldspar)

Gneiss
Foliation dips 70° to NE

(a) (i) Draw the large-scale folds which are present in the area. What evidence in the rock faces can be used to work out the shapes of the folds?

   (ii) Fig. 8.4B shows thin sections of two samples of one rock type from rock face B. Identify the rock and account for the differences between the specimens.

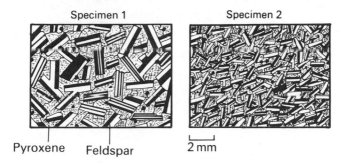

Specimen 1   Specimen 2

Fig. 8.4B   Pyroxene   Feldspar   └─┘ 2 mm

(b) Radiometric dating of rock samples gave these results:

shale with *Monograptus* (rock face A): 430 ± 10 Ma
shale with ammonites (rock face B): 175 ± 5 Ma
igneous rock (rock face B): 375 ± 10 Ma
pegmatite (rock face D): 1200 ± 30 Ma
gneiss (rock face D): 1200 ± 30 Ma

At rock face D, the pegmatite cuts the gneiss so it seems to post-date the gneiss. However, the rocks have identical radiometric ages. Explain this apparent anomaly.

(c) The topmost limestones and shales have distinctive ammonite species distributed as follows:

	Rock face A	Rock face B	Rock face C	Rock face D
Youngest	Species 4	Species 4	Species 4	Species 4
↑	Species 3	Species 3	Species 4	Species 4
	Species 2	Species 2	Species 3	Species 4
Oldest	Species 1	Species 2	Species 3	Species 4

Describe, with diagrams, the pattern of sedimentation which could have produced this distribution of fossils.

(d) Which three of the statements A–H are correct?

A The igneous rock at rock face B is a dyke because it is nearly vertical.

B At rock face A, the limestones immediately above the greywackes and shales are of Carboniferous age.

C At rock face B, the uppermost shales and limestones are of Triassic age.

D The greywackes and shales are of Silurian age.

E The cleavage in the shale could be the same age as the foliation in the gneiss.

F The unconformity in rock face C cannot be the same as the unconformity in rock face D.

G   There is a pre-Jurassic fault between rock faces C and D which throws down to the south-west.
H   The currents which deposited the sand at the top of rock face D flowed from the north-east towards the south-west.

(e) Pollen types found at different levels in the peat above rock face B are as follows:

VI (*youngest*)   Alder, oak, birch, hazel, beech
V   Conifers, beech, hornbeam
IV   Oak, lime
III   Oak, hazel
II   Scots pine, birch, elm, spruce
I (*oldest*)   Dwarf birch, shrubby willows, mountain avens

Describe in a very general way how the climate changed during the period of peat deposition.

8.5   Fig. 8.5 shows faulted and intruded sedimentary rocks. How can you tell that the following statements are true?

(a) Igneous rocks I and J are parts of different intrusions.
(b) Igneous rock K forms a sill, not a lava flow.
(c) Dyke D is older than fault F.
(d) Fault F is a tear fault.
(e) Fault F is older than intrusion J.

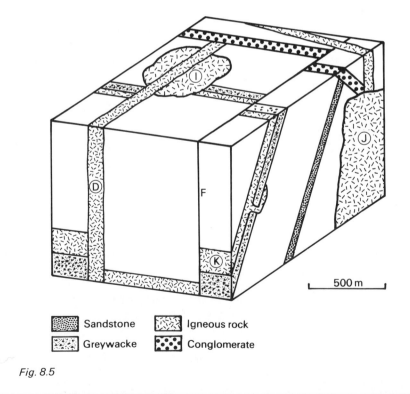

Sandstone   Igneous rock
Greywacke   Conglomerate

Fig. 8.5

8.6 Study the map (Fig. 8.6A).
  (a) Place faults H, J, K and L in their correct order from old to young.
  (b) Between which two geological periods was fault M formed?

Fig. 8.6A

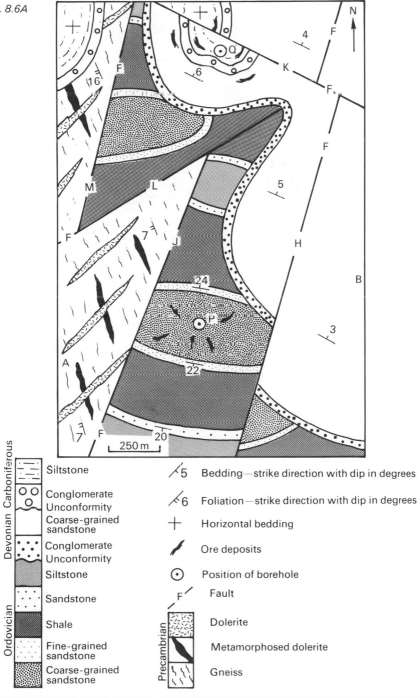

Devonian Carboniferous

	Siltstone
	Conglomerate Unconformity
	Coarse-grained sandstone

Ordovician

	Conglomerate Unconformity
	Siltstone
	Sandstone
	Shale
	Fine-grained sandstone
	Coarse-grained sandstone

⋏5 Bedding—strike direction with dip in degrees

⋏6 Foliation—strike direction with dip in degrees

+   Horizontal bedding

⨍   Ore deposits

⊙   Position of borehole

F⫽ Fault

Precambrian

	Dolerite
	Metamorphosed dolerite
	Gneiss

(c) Boreholes at P and Q passed through the rocks shown in Fig. 8.6B. Which statement correctly describes relationships shown by the ore deposits?

A In borehole P the ore is of Ordovician age. In borehole Q some of the ore is Devonian and some is Carboniferous.

B In borehole P the ore could have been formed by fluids from the granite. In borehole Q the ores have not been formed by fluids from the granite.

C In borehole P the ore has been formed by hydrothermal processes. The ore in borehole Q has been formed by sedimentary processes.

D In both boreholes the ores have probably been formed by hydrothermal fluids from the underlying granites.

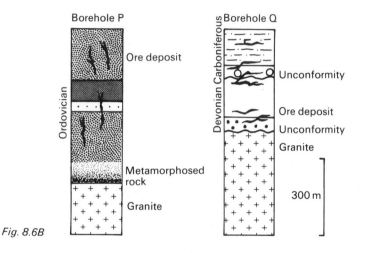

Fig. 8.6B

(d) The amount of heat escaping from the ground surface was measured along a line from A to B on the map. The form of the ground surface and the corresponding heat flow are shown in Fig. 8.6C.

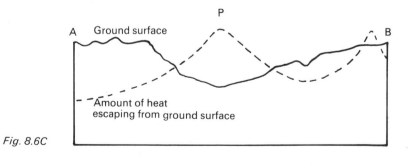

Fig. 8.6C

(i) Where on the map are additional, unmarked ore deposits likely to be found? Give a reason for your answer.

(ii) Is the granite in borehole P older or younger than fault J? Is the granite older or younger than fault H? Give a reason for your answer.

(e) From the following list choose the two statements which correctly describe the relationships on the map.
A Fault M is the same age as fault H.
B Fault J could be a normal fault throwing down to the west.
C The metamorphosed dolerite in the Precambrian was metamorphosed before the dykes were intruded.
D The metamorphosed dolerite in the Precambrian was metamorphosed during the intrusion of the dykes.
E The Devonian is separated from the Ordovician by a thrust fault.
F The Carboniferous is conformable on the Devonian but unconformable on the Precambrian.
G Fault J could be a normal fault throwing down to the east.
H The ore deposits at Q were formed later than faulting on K.
(f) Write a brief geological history of the area shown in the map.

8.7 A constantly-dipping coal seam was found at the depths shown in three boreholes (Fig. 8.7).

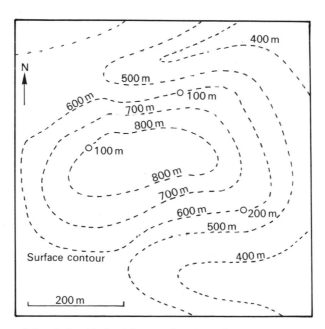

O Borehole with depth from surface to coal seam
200 m

Fig. 8.7

Make a tracing of the map.
(a) Draw structure contours (strike lines) for the coal seam.
(b) What is the angle and direction of dip of the seam?
(c) Draw in the outcrop of the seam.

8.8 On the map, Fig. 8.8, a constantly-dipping coal seam crops out along CD.

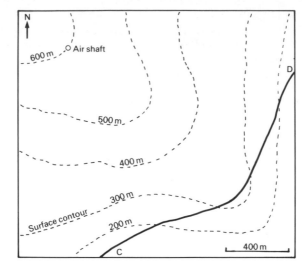

Fig. 8.8

Make a tracing of the map.
- (a) Find the direction and amount of dip of the coal seam.
- (b) Draw in the outcrop (EF) of a second seam 100 m above seam CD in the sedimentary sequence.
   Note: the distance is measured at right angles to CD.
- (c) A colliery manager has decided to sink an air shaft from a point at 600 m OD. At what depths will the air shaft meet seams EF and CD?
- (d) On the map, shade the area which consists of rocks older than seam CD.

8.9 Make a tracing of the map, Fig. 8.9. Fault FF throws down 100 m to the west.

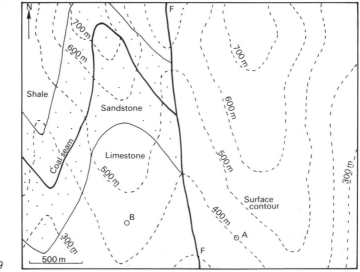

Fig. 8.9

(a) Find the direction and amount of dip of the coal seam.
(b) Complete the outcrops to the east of the fault.
(c) At what depths will the coal seam be found in boreholes A and B?
(d) How thick is the sandstone which holds the coal seam?

8.10 On the map, Fig. 8.10, fault FF throws down 200 m to the east. The base of a bed of sandstone outcrops at A, B and C. The base of the sandstone is 200 m vertically below the top of the sandstone.

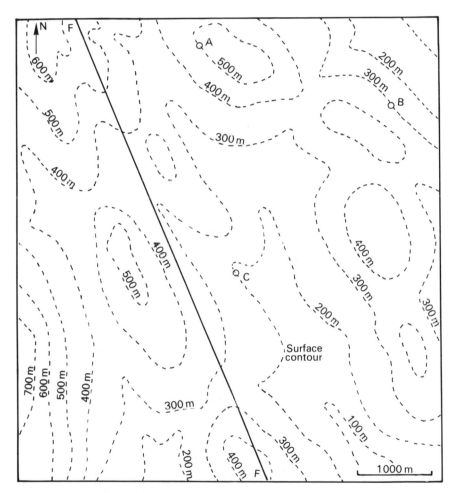

Fig. 8.10

Make a tracing of the map.
(a) Find the direction and amount of dip of the sandstone.
(b) Draw in the outcrop of the sandstone on both sides of the fault.
(c) How can you tell that the fault plane is vertical?

8.11 On the map, Fig. 8.11, points A, B and X lie on outcrops of a constantly-dipping coal seam. Fault FF has a vertical displacement. A borehole at C meets the coal seam at a depth of 300 m from the surface. Trace the map.

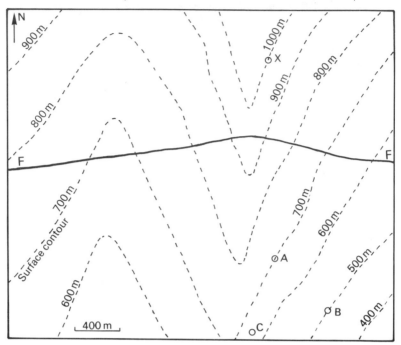

Fig. 8.11

(a) Find the direction and amount of dip of the coal seam.
(b) Draw in the outcrop of the seam on both sides of the fault.
(c) Find the direction and amount of throw of the fault.
(d) Why is the outcrop of the fault plane not a straight line?

8.12 Identify the features shown in the photograph and say how they were formed.

8.13 The satellite photograph on this book's back cover shows part of northern Scotland. Skye is just left of centre. Study a map of the regional geology of the area and relate the geological features to the topographic features shown in the photograph.

8.14 Fig. 8.14 shows an area of continental and oceanic crust. The arrows show the directions and speeds (in cm per year) of plate movement. Make a tracing of the map.

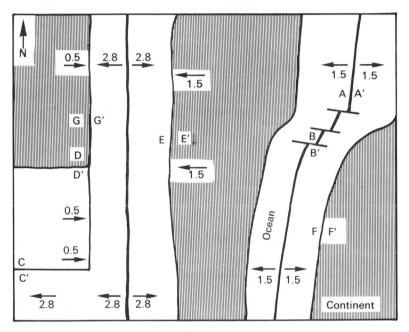

Fig. 8.14

(a) How many plates are shown? Number the plates from west to east.
(b) Label the destructive, constructive and conservative plate boundaries.
(c) Show the following features using the symbols given below.

    Young mountains of Andean type   M–M–M–M
    Shallow-focus earthquakes   S–S–S–S
    Deep-focus earthquakes   D–D–D–D
    Oceanic trench   T–T–T–T
    Oceanic ridge   R–R–R–R
    Island arc   I–I–I–I

(d) Where would you find the following?
Area of youngest crust; areas of high heat flow; area of strong negative gravity anomaly; explosive volcanic activity; effusive volcanic activity; pillow lavas.
(e) What are the relative speeds and directions of plate movement between the pairs of points A–A' to G–G'?

8.15 Fig. 8.15 gives details of four continents.

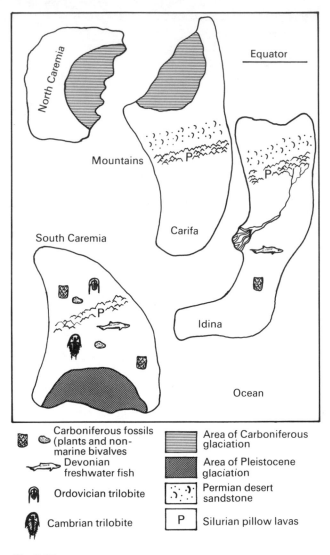

Fig. 8.15

(a) Give four pieces of evidence which suggest that these continents once formed a single land mass.
(b) The single land mass is thought to have formed by the coming together of still earlier land masses. Give two pieces of evidence which support this suggestion.

GEOLOGICAL TIME

9.1 (a) A delta has a thickness of 160 m and an area of 35 200 km². What is the volume of the delta?

(b) Throughout its history, 1.05×10^8 m³ of sediment have been added to the delta every year. What is the age of the delta?

(c) Using the same reasoning, Charles Lyell worked out the age of the Mississippi Delta. How was Lyell able to conclude: 'Yet the whole period during which the Mississippi has been transporting its earthy burden to the ocean, though perhaps far exceeding 100 000 years, must be insignificant in a geological point of view, . . .'?

9.2 By using measurements of heat flow and by assuming that the Earth was originally molten, Lord Kelvin estimated the age of the Earth to be between 20 and 400 Ma old, with 100 Ma being the most probable age. Later, he refined his estimate to 20–40 Ma.

(a) Why did Kelvin's estimates vary so much?

(b) Why are Kelvin's estimates much less than present-day estimates of the age of the Earth?

9.3 (a) When a radioactive parent isotope decays, it produces a stable daughter isotope which does not decay. What is the *half-life* of a radioactive isotope?

(b) If we begin with a certain number of original atoms which decay over four half-lives to leave a number of surviving atoms, we can say:

$$\text{survivors} = \text{originals} \times \tfrac{1}{2} \times \tfrac{1}{2} \times \tfrac{1}{2} \times \tfrac{1}{2}$$

$$N_s = N_o \times (\tfrac{1}{2})^4$$

More generally, for n half-lives: $N_s = N_o \times (\tfrac{1}{2})^n$

The ratio of surviving to original atoms is: $\dfrac{N_s}{N_o} = (\tfrac{1}{2})^n$

Taking logs: $\log\left(\dfrac{N_s}{N_o}\right) = n \log(\tfrac{1}{2})$

And the number of half-lives is: $n = \dfrac{\log\left(\dfrac{N_s}{N_o}\right)}{\log(\tfrac{1}{2})}$

(i) In a mineral sample, the ratio of parent to daughter atoms is 1 : 12. The parent isotope has a half-life of 5 million years (Ma). What is the age of the sample?

(ii) In a sample of zircon the ratio of ^{238}U atoms to ^{206}Pb atoms is $1:0.4$. The half-life of ^{238}U is 4500 Ma. What is the age of the sample?

(iii) In the Earth as a whole, U–Pb atomic ratios are:

$$\frac{^{238}U}{^{206}Pb} = 0.5405$$

$$\frac{^{235}U}{^{207}Pb} = 0.0046$$

Using these ratios, find the age of the Earth. (The half-life of ^{235}U is 710 Ma.)

(iv) Why did you get two different answers to part (iii)?

(v) Analysis of meteorites gives a good indication of how much ^{206}Pb and ^{207}Pb were present in the Earth when it was formed. Subtracting these values from present Earth levels gives these modified U–Pb ratios:

$$\frac{^{238}U}{^{206}Pb} = 0.9524$$

$$\frac{^{235}U}{^{207}Pb} = 0.011$$

Find the age of the Earth using these ratios. How do your answers compare with those found in part (iii)?

(c) Radiometric ages are given in the form 3670 ± 90 Ma. What is the significance of the $\pm$ number?

(d) In addition to analytical error, state two sources of error which may arise in determining radiometric ages.

9.4 (a) (i) Every year, rivers carry 6×10^7 tonnes of sodium to the sea. The oceans contain about 15×10^{15} t sodium. Assuming that sodium has been accumulating throughout geological time, how old are the oceans?

(ii) Would annual sodium input from rivers to the sea necessarily have been the same throughout geological time? Explain your answer.

(iii) Why is the figure obtained in part (i) much less than the true age of the oceans?

(iv) In the sodium cycle (Fig. 9.4), which of the following phrases are represented by letters A–F?
Chemical weathering; sodium in igneous rocks; sodium in metamorphic rocks; sodium in rain and wind-blown spray; sodium in sedimentary rocks; transport by rivers.

(b) The residence time of an element in the oceans is the total amount of the element in the oceans divided by the amount introduced annually. This means that, on average, an atom of the element spends this length of time in the ocean before being moved to another stage in its geochemical cycle. What is the residence time of sodium? Why does sodium have a much longer residence time than calcium (residence time: 8.0×10^6 years), silicon (residence time: 8×10^3 years) and manganese (1.4×10^3 years)?

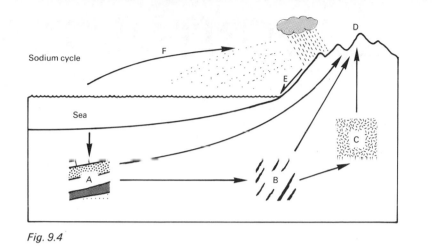

Sodium cycle

F

Sea

A   B   C   D   E

Fig. 9.4

9.5 The Lizard Complex of Cornwall may represent a thrust mass carried from the south during the Hercynian (Variscan) Orogeny.

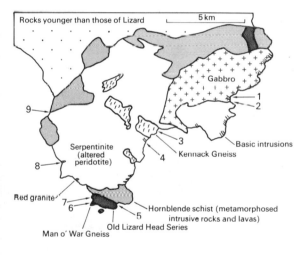

Rocks younger than those of Lizard |⎯ 5 km ⎯|

Gabbro

9

Serpentinite
(altered
peridotite)

8

Red granite

7
6

3
Kennack Gneiss
4

Basic intrusions

Hornblende schist (metamorphosed
intrusive rocks and lavas)

5
Old Lizard Head Series

Man o' War Gneiss

1 Gabbro dyke cut by olivine dolerite dyke
2 Serpentinite cut by olivine gabbro cut by gabbro
3 Kennack Gneiss cut by red granite
4 Serpentinite cut by gabbro cut by dolerite cut by Kennack Gneiss
5 Interlayering of hornblende schist and Old Lizard Head Series
6 Old Lizard Head Series intruded by Man o'War Gneiss
7 Old Lizard Head Series metamorphosed by Man o'War Gneiss then by serpentinite
8 Serpentinite cut by red granite
9 Hornblende schist cut by serpentinite

Fig. 9.5

(a) From the information in the map, Fig. 9.5, place the rock types into their correct order from old to young.

(b) It is thought that part of the Lizard Complex is late Precambrian basement. The last phase of igneous activity and metamorphism took place 350 Ma ago. Why do none of the rocks give Precambrian potassium–argon ages? Why do some of the hornblende schists (which appear to be Precambrian) give potassium–argon ages of about 450 Ma?

PALAEONTOLOGY

10.1   (a) The skeletons of modern corals consist of aragonite. What is aragonite?

     (b) Under what conditions do modern coral reefs grow?

     (c) The cells of coral polyps contain mutualistic algae. What kind of ecological relationship is mutualism? How do the algae benefit the coral?

     (d) Account for the following observations made on a modern reef.

        1 Corals grow towards light.

        2 In shallow water, corals have robust skeletons.

        3 Corals growing in the shade of others may produce long, tree-like branches.

        4 Polyps in deep water contain more algae than polyps in shallow water.

        5 The growth of coral is much faster in light than in darkness. The rate at which aragonite is deposited decreases with depth.

     (e) Corals are carnivorous but they can also take in dissolved and particulate organic matter. Their mutualistic algae also produce food. What is the advantage of having various methods of obtaining food?

     (f) What processes cause reefs to lose carbonate material?

     (g) Fig. 10.1 shows a section through a coral reef.

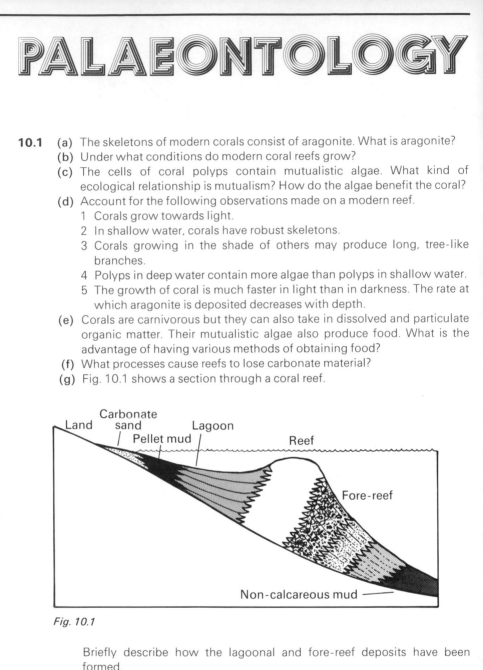

Fig. 10.1

        Briefly describe how the lagoonal and fore-reef deposits have been formed.

     (h) Name four organisms which have been involved in reef building since Cambrian times.

10.2   (a) What is a species?

     (b) What criteria would you use to decide if fossils belonged to the same or different species?

(c) Single shells from disarticulated bivalves were collected on a beach and their heights and lengths measured (Table 10.2A). Plot a scattergram of height against length.

How many species of bivalve appear to be present?

Shell	Height	Length	Shell	Height	Length
1	2.37	5.39	41	1.71	3.32
2	2.35	4.91	42	2.08	3.82
3	2.80	5.78	43	1.72	3.44
4	2.49	4.56	44	2.25	3.84
5	2.62	4.82	45	2.15	3.73
6	2.17	4.73	46	2.15	3.78
7	2.17	4.80	47	1.98	3.72
8	2.54	4.23	48	1.92	3.78
9	2.21	4.31	49	2.02	3.63
10	2.15	4.11	50	1.97	3.56
11	2.32	4.99	51	2.02	3.47
12	2.20	4.66	52	1.91	3.56
13	2.08	4.08	53	1.87	2.98
14	1.47	2.39	54	1.73	3.17
15	1.51	2.77	55	1.89	3.32
16	1.34	2.47	56	1.76	3.53
17	1.72	2.58	57	1.87	3.44
18	1.79	3.09	58	1.78	3.26
19	1.71	2.93	59	1.64	3.31
20	1.44	2.64	60	2.06	3.19
21	1.71	2.81	61	3.10	3.99
22	1.76	3.36	62	3.05	3.87
23	1.51	2.57	63	2.76	3.53
24	0.83	1.48	64	1.99	2.69
25	1.48	2.65	65	3.05	3.61
26	1.59	2.71	66	2.16	2.75
27	1.48	2.58	67	2.66	3.45
28	1.39	3.00	68	3.07	3.96
29	1.32	2.19	69	2.78	3.69
30	1.41	2.70	70	2.93	3.71
31	1.32	2.50	71	2.71	3.28
32	1.48	2.85	72	1.61	2.22
33	1.06	1.84	73	1.53	2.11
34	1.37	2.82	74	2.41	3.21
35	1.53	2.70	75	2.30	3.02
36	1.59	2.76	76	2.34	3.03
37	1.76	2.84	77	1.59	2.29
38	2.09	4.00	78	1.22	1.71
39	2.41	4.80	79	1.19	1.58
40	1.91	3.64			

Table 10.2A: all measurements in cm

(d) Shells with similar height/width ratios do not necessarily look the same. Draw two bivalves with similar ratios but different shapes.

(e) Samples of banded carpet shell (*Venerupis*) were also collected (Table 10.2B). Plot the measurements for these shells onto the scattergram drawn in part (c).

Shell	Height	Length
1	2.08	2.92
2	2.02	3.10
3	1.73	2.60
4	2.04	2.83
5	1.84	2.78
6	1.91	2.99
7	2.09	3.02
8	2.09	2.94
9	1.53	2.48
10	1.32	2.18
11	1.86	2.80
12	2.00	2.93
13	2.19	3.04
14	1.65	2.54
15	2.17	3.16
16	1.29	1.94
17	1.66	2.51

Table 10.2B: Measurements (in cm) made on banded carpet shell (Venerupis).

Would measurements of height and length alone allow separation of all the collected shells into species? If not, what other measurements or shell characteristics might be taken into account?

10.3 Fig. 10.3 shows the geological ranges of some brachiopod orders. The thicker the column, the more abundant the members of the order.

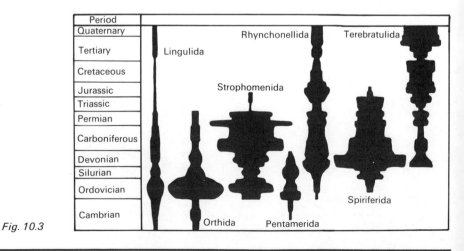

Fig. 10.3

(a) How does biological variation differ from diversity? During which periods were brachiopods most diverse?

(b) Give two ways in which articulate brachiopods differ from inarticulate brachiopods. Which orders in the diagram belong to the class Inarticulata?

(c) Give two reasons why orders may have become extinct and two reasons why orders may have expanded.

(d) What is a faunal province?

(e) During the early Ordovician, southern British brachiopods belonged to the European faunal province while northern brachiopods belonged to the American province. Why were the brachiopod faunas no longer provincial towards the end of the Ordovician?

10.4 (a) In the *Nautilus* shell, what are camerae, septa, suture lines, siphuncle and body chamber? What are the functions of the camerae, septa and siphuncle?

(b) How do ammonoid septa and suture lines differ from those of nautiloids? What is the advantage to the ammonoids of having septa of the type described?

(c) *Nautilus* shells are crushed by pressures of 6.0×10^6 N m^{-2}. What is the greatest depth at which *Nautilus* can live? (Sea water has a density of 1.025×10^3 kg m^{-3}.)

(d) At depth, the blood inside the siphuncle of *Nautilus* is at very high pressure while the gas in the camerae is at very low pressure. What is the function of the calcareous tube around the siphuncle?

(e) Ammonoids have no calcareous walls on their siphuncles. What does this suggest about the depths at which they lived?

(f) The centre of gravity or centre of mass of a body is the point through which the weight may be considered as acting. In a floating body, the centre of buoyancy is the centre of gravity of the displaced water. A sealed floating test-tube with most of its volume submerged has lead shot in its base. Is its centre of gravity above, below or in the same place as its centre of buoyancy? If the test-tube is tilted then released, how readily will it return to its original floating position?

(g) To swim, the *Nautilus* blows water from its hyponome (see Fig. 10.4). Why does the animal not spin like a wheel?

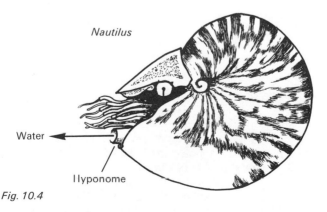

Nautilus

Water

Hyponome

Fig. 10.4

(h) Some ammonoids had centres of buoyancy and centres of gravity in about the same place. What does this suggest about their swimming ability?

(i) Late Cretaceous ammonitids were commonly of two forms:

1 Shapes with broad cross-sections, poorly adapted for swimming. Strong ribs, spines and tubercles present.

2 Streamlined shapes with narrow cross-sections, well adapted for swimming. No ribs, spines or tubercles.

How were these characteristics advantageous to the two forms?

10.5 (a) What is an adaptation?

(b) How is a burrowing sea urchin such as the sea potato (*Echinocardium*) adapted to its habitat? How can knowledge obtained from study of living forms such as this help us to interpret the way in which extinct forms (e.g. *Micraster*) lived?

(c) Briefly describe the modes of life of the following bivalves.

Mussel (*Mytilus*); oyster (*Ostrea*); scallop (*Pecten*); cockle (*Cerastoderma*); razor-shell (*Ensis* or *Solen*); gaper (*Mya*); piddock (*Pholas*).

How are the shell forms adapted to the ways in which the animals live?

10.6 Dog whelks are carnivorous gastropods which live on rocky shores. Dog whelks were collected from positions 1 and 2 in Fig. 10.6A and the living contents were removed. Dead shells were collected from the beach at position 3 and from the raised beach deposits at position 4. Measured properties of the empty shells are given in the Table 10.6. Shells from positions 1 and 2 are shown in Fig. 10.6B.

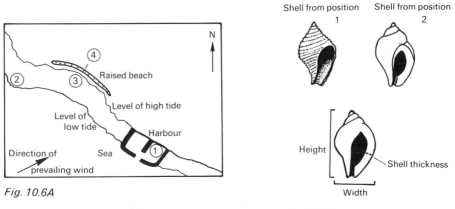

Fig. 10.6A

Fig. 10.6B

(a) When the dead dog whelks are washed from the rocks, in which direction will most shells be transported? Give a reason for your answer.

(b) Explain how the differences have arisen between the shells at position 2 and the shells at the other positions.

(c) Explain why the mean values of height, width, thickness and weight are greater at positions 3 and 4 than at positions 1 and 2.

	Position			
	1	2	3	4
Number in sample	67	95	59	20
Mean height (cm)	2.06	2.17	2.42	2.43
Sample standard deviation (SD)	0.47	0.53	0.27	0.24
Mean width (cm)	1.26	1.38	1.59	1.56
Sample SD	0.27	0.33	0.16	0.12
Mean height/width ratio	1.00	1.57	1.52	1.55
Sample SD	0.08	0.08	0.08	0.10
Mean thickness (cm)	0.15	0.23	0.25	0.28
Sample SD	0.07	0.08	0.06	0.05
Mean weight (g)	1.43	2.30	3.20	3.35
Sample SD	1.25	1.42	1.06	0.93
Number of shells in each height range (cm)				
0.8–1.19	1	3		
1.2–1.59	8	15		
1.6–1.99	26	12	3	
2.0–2.39	19	25	23	7
2.4–2.79	8	30	29	12
2.8–3.19	3	10	4	1
3.2–3.69	2			
Number of shells in each width range (cm)				
0.4–0.59		1		
0.6–0.79	2	3		
0.8–0.99	6	13		
1.0–1.19	25	10		
1.2–1.39	14	13	6	1
1.4–1.59	12	22	25	10
1.6–1.79	3	29	23	9
1.8–1.99	5	4	5	
Number of shells in each range of height/width ratio				
1.3–1.39			3	
1.4–1.49	3	18	18	6
1.5–1.59	16	49	27	8
1.6–1.69	36	22	10	4
1.7–1.79	10	4	1	1
1.8–1.89	2	2		1
Number of shells in each thickness range (cm)				
0.0–0.09	5	1		
0.1–0.19	54	40	14	
0.2–0.29	4	29	30	12
0.3–0.39	3	25	15	8
0.4–0.49	1			
Number of shells in each weight range (g)				
0.0–0.99	33	28		
1.0–1.99	25	14	8	2
2.0–2.99	4	15	21	5
3.0–3.99	0	26	17	9
4.0–4.99	3	10	9	3
5.0–5.99	1	2	4	1
6.0–6.99	1			

Table 10.6

(d)   Explain why the values of standard deviation for height, width, thickness and weight are lower at positions 3 and 4 than at positions 1 and 2.

(e)   Explain why the height/width ratio of the shells at positions 1 and 2 is greater than that of the shells at positions 3 and 4.

(f)   Despite a long search of the raised beach deposits, only 20 shells were found. What does this suggest about the chances of preservation of dog whelk shells?

(g)   When sand from position 3 was added to dilute hydrochloric acid, 90% of it dissolved. What is the probable source of the sand?

(h)   The raised beach deposits are thought to be about 5 000 years old. The shells are heavy, thick and round. Such shell properties would appear to be adaptations to life in an exposed habitat. Does this indicate that the seas 5 000 years ago were stormier than seas in this area today? Explain your answer.

(i)   If the samples from positions 1 and 2 were combined, would this combined sample be representative of the dog whelks of the whole coast? Give a reason for your answer.

(j)   At position 1 most of the shells are thin and light. A few shells are thick and heavy. Explain how these thick, heavy shells have come to be at this position.

(k)   How may a fossil community differ from the living community from which it was derived?

(l)   How may a fossil population differ from the living population from which it was derived?

(m)   How does a life assemblage differ from a death assemblage? Do the dog whelks in the raised beach deposits form a life assemblage or a death assemblage?

(n)   When studying the properties of fossil populations, state one disadvantage of having a small sample.

10.7   In the photograph, identify the features of biological origin. How were they formed?

EARTH HISTORY

11.1 (a) Classify the following stratigraphical units into time units, time-stratigraphical units and rock units.
Bed; era; formation; group; period; series; stage; system.

(b) What is a zone? What properties should be possessed by a good zone fossil?

(c) What are facies fossils?

(d) Why is correlation difficult in the Devonian of Britain, but relatively easy in the Jurassic?

11.2 In north-east Scotland, many of the Devonian rocks were deposited in tropical lakes. Every year, a hot, dry season was followed by a warm, wet season. Dense algal blooms grew during the dry season.

Deposits of three main types can be distinguished.

Type I: Varved sediments deposited in deep, stratified lakes (Fig. 11.2). Fish fossils are numerous. Many are complete. Fish scales have well developed rings.

Type II: Carbon-rich shales and siltstones deposited in shallow, unstratified lakes. Mudcracks are present. Fish fossils are rare. The fossils are disarticulated.

Type III: Shales and siltstones deposited in shallow, unstratified lakes. Mudcracks are common. Halite is present. Fish fossils are very rare. The fossils are disarticulated.

Lake varves

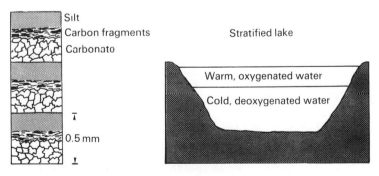

Fig. 11.2

(a) Why do the fish scales have well-developed rings?

(b) What would be the effects on the lakes of:
1 Several years of heavy rainfall?
2 Several years of low rainfall?
How would these changes affect fish distribution?

(c) Suggest five factors which would influence the size of a fish population.

(d) Within what period of time is a varve deposited?

(e) During which seasons were the carbonate and silt units of the varves deposited?

(f) During the dry season, planktonic algae grew rapidly and their photosynthetic activity removed CO_2 from the water. How would the reduction of CO_2 cause carbonate precipitation in the varves?

(g) In the varved sediment, what is the source of the small carbon fragments? Why do the carbon fragments appear above the carbonate?

(h) In which part of a stratified lake would fish live?

(i) Why are fish fossils common and often complete in the deposits formed in the stratified lakes?

(j) If a dead fish has a width of 5 cm, how long would it take for it to be buried by varved sediment?

(k) What effects would evaporative water loss have on fish populations?

(l) Why are fish fossils rare and disarticulated in deposits of Types II and III?

(m) What are coprolites and what do they indicate about fish diets?

(n) Suggest possible food webs for the lake communities. (There are no fossils of zooplankton but they may have been present.) Are your food webs simpler or more complex than those of modern lakes?

(o) Why is correlation difficult among lake sequences of this type?

(p) There is one Caithness fish bed which can be correlated with beds in Orkney, Shetland and around the Moray Firth. What characteristics might make such a bed distinctive?

11.3 From north to south over a distance of 200 km, an area of Permian rocks showed the following contemporaneous sedimentary facies:

1 Red finely-laminated siltstones and mudstones. Mudcracks are common.

2 Interlaminated gypsum and dolomite.

3 Dolomites with numerous faecal pellets and stromatolites.

4 Massive limestone with very numerous fossils of bryozoans, calcareous algae, sponges, brachiopods, crinoids and foraminiferans.

5 Limestone breccias and boulder beds deposited on slopes inclined at 20° towards the south. Fossils include delicate bryozoans.

6 Black shales and siltstones with a few fine-grained limestones. Fossils include fish, ammonoids, foraminiferans and radiolarians. Sandstone showing graded bedding forms channels through the shales. The sandstone channels run north–south.

(a) Under what depositional environments were the various sedimentary facies formed?

(b) Briefly describe the overall pattern of palaeogeography and sedimentation.

11.4 Fig. 11.4 shows the palaeogeography of the British area during the Ordovician Period.

(a) What are ophiolites?

(b) What evidence suggests that subduction took place under the Southern Uplands and under the Lake District? Does any evidence remain of a constructive plate margin?

(c) Under what conditions was the Ordovician of North Wales deposited?

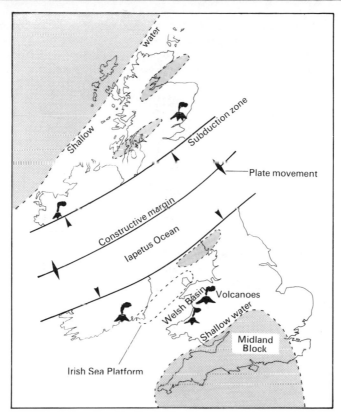

Fig. 11.4

(d) Give a very brief outline of the tectonic events which accompanied closure of the Iapetus Ocean.

11.5 (a) Briefly describe the pattern of sedimentation which produced the sequences shown in Fig. 11.5A. What term is used to describe a thin but complete sequence like that at position 2?

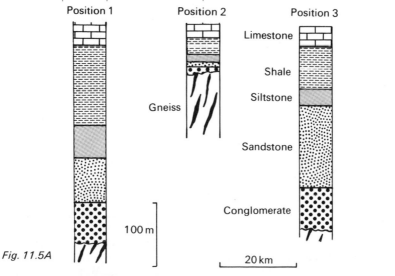

Fig. 11.5A

(b) (i) Briefly describe the pattern of sedimentation which produced the sequences shown in Fig. 11.5B.

(ii) Does the direction of thickening of the deltaic sandstone allow the position of an adjacent land mass to be determined? Explain your answer.

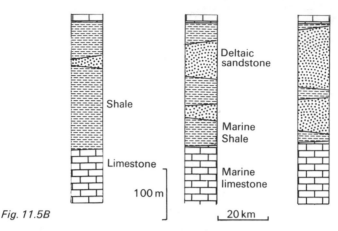

Fig. 11.5B

(c) Thickness variation in a sandstone and conglomerate rock body have been found by close-spaced drilling. On a tracing of Fig. 11.5C, draw an isopachyte map for the rock.

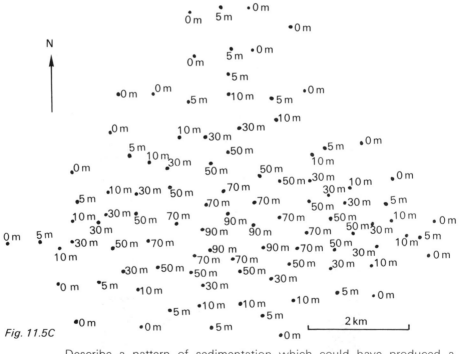

Fig. 11.5C

Describe a pattern of sedimentation which could have produced a depositional body of this type.

11.6 Fig. 11.6 shows the palaeogeography of the British area during the Permian Period.

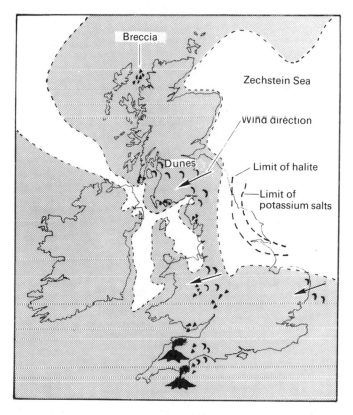

Fig. 11.6

(a) What is brockram and how has it been formed?

(b) Some Permian sandstones consist of millet-seed sand. How is millet seed sand formed?

(c) How have palaeowind directions been determined?

(d) Why do the limits of halite and potassium salts differ?

11.7 Fig. 11.7 shows the palaeogeography of the British area during the Tertiary
Period.

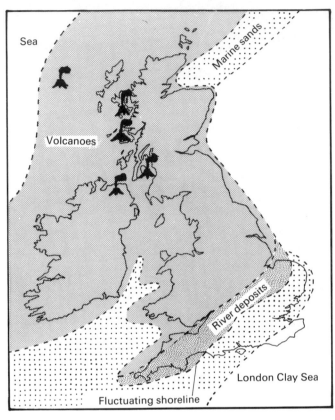

Fig. 11.7

(a) What evidence from the early Tertiary rocks of the London and Hampshire
Basins suggests that the position of the shoreline fluctuated?
(b) What evidence is used to identify the sediments as marine, estuarine or
fluvial?
(c) What evidence is used to determine Tertiary climatic conditions?

EARTH RESOURCES

12.1 (a) One way of measuring the permeability of coarse-grained material is to use a constant head permeameter (Fig. 12.1A).

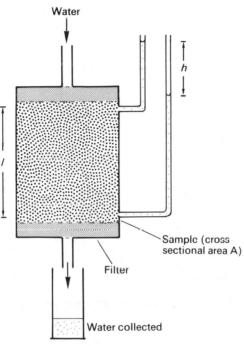

Fig. 12.1A

(i) Why must undisturbed samples be used?
(ii) Why must the water be vacuum treated before use?
(iii) Why are filters inserted?
(iv) What is hydraulic gradient? In which units is it measured?
(v) Why does the pressure drop across the filters not affect the measurement of permeability?

(b) The coefficient of permeability (k) is: $k = \dfrac{ql}{Ah}$

where q is the flow of water ($m^3\,s^{-1}$); l is the vertical distance (m) between the side tubes; A is the cross sectional area (m^2) of the sample; and h is the difference in height (m) between the water levels in the side tubes.

(i) In which units is the coefficient of permeability measured?
(ii) The following measurements were made on a sample of sand:

$q = 7.5 \times 10^{-6}\,m^3\,s^{-1}$; $A = 0.01\,m^2$; $l = 0.2\,m$; $h = 0.15\,m$

What is the coefficient of permeability of the sand?

(iii) The following measurements were made on a sample of gravel:

$q = 2.5 \times 10^{-4}\,\mathrm{m^3\,s^{-1}}; A = 0.01\,\mathrm{m^2}; l = 0.2\,\mathrm{m}; h = 0.05\,\mathrm{m}$

What is the permeability of the gravel?

(c) (i) What property should be possessed by rock underlying a site suitable for the disposal of solid and liquid toxic wastes?

(ii) How should such a site be treated after filling?

(d) Fig. 12.1B shows a disposal site for toxic wastes. Water flows through the rocks and refuse at the following rates:

Material	Rate of flow $(\mathrm{m\,yr^{-1}})$
Refuse	200
Clay	5×10^{-6}
Siltstone	1.10
Sand	41
Limestone	70

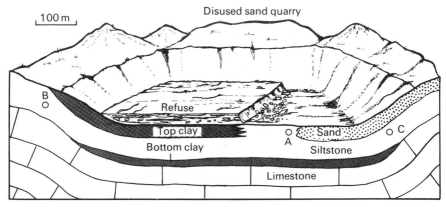

Fig. 12.1B

(i) How should the site have been treated before dumping began?

(ii) Why is the refuse much more permeable than the other materials?

(iii) The bottom clay is 1.6 m thick and the thickest part of the siltstone is 4.2 m thick. What would be the shortest time taken for a toxic substance to reach the limestone?

(iv) What is the major disadvantage of the site?

(v) What would be the shortest times taken for toxic materials dissolved in ground water to travel from point A to points B and C in Fig. 12.1 B? Why is the shortest time not necessarily the actual time?

(vi) By means of a diagram, show the likely directions of groundwater flow in the siltstone and limestone.

12.2 Complete the key in Fig. 12.2 using the words and phrases listed below.
Banded iron formations; contact metasomatism; cumulates; diamonds in
kimberlite; exhaled hydrothermal solutions; liquid immiscibility; mechanical
accumulation; metamorphic processes; lead–zinc of south Pennines;
pegmatites; residual processes; secondary or supergene enrichment;
tin–copper–tungsten of Cornwall.

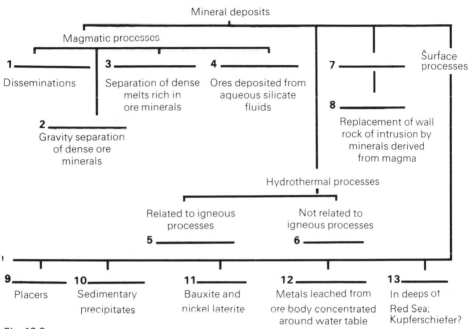

Fig. 12.2

12.3 (a) (i) What is a geochemical anomaly? Why are such anomalies often
associated with ore bodies?
(ii) Anomalous concentrations of elements may not lie directly above
ore bodies. On copies of Figs. 12.3A–12.3C, show how elements
from the ore bodies may be dispersed by surface processes.

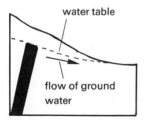

dispersion by ground
water

Fig. 12.3A

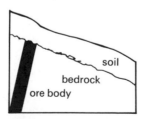

dispersion by weathering and
soil creep

Fig. 12.3B

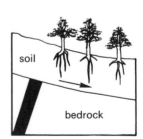

dispersion by plants

Fig. 12.3C

(iii) Fig. 12.3D shows the geochemical anomaly associated with a copper deposit at Coed-y-Brenin, North Wales. Explain why most of the anomaly does not overlie the ore bodies.

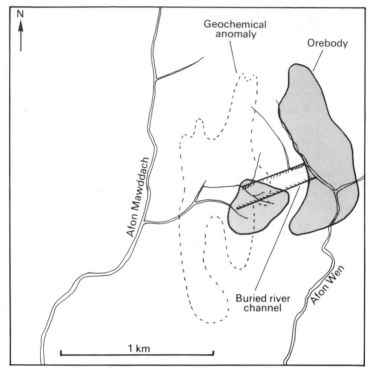

Fig. 12.3D

(b) Magnetic surveys are often used to find ore bodies. Some values of the relative magnetic strengths of geological materials are shown in Table 12.3.

Material	Relative strength of magnetism
Chalcopyrite	0.4
Fluorite	0.0
Gold	0.0
Haematite	7.0
Ilmenite	1880.0
Sphalerite	0.8
Alluvium	0.0
Diorite	88.0
Granite	3.0
Limestone	0.3

Table 12.3

How easily could the following deposits be detected by magnetic survey?
(i) A placer deposit rich in gold.
(ii) A placer deposit rich in ilmenite.
(iii) A thick fluorite vein in limestone.
(iv) Diorite rich in chalcopyrite intruded into limestone.
(v) A thick haematite-rich vein in limestone.
(vi) A thick sphalerite-rich vein in granite.

12.4 **(a)** (i) What is the calorific value of a fuel?
(ii) What is coalification?
(iii) What is the rank of a coal? What processes contribute to an increase in rank?
(b) Table 12.4 shows the average composition in weight per cent of wood, peat and coals.

	Carbon	Hydrogen	Nitrogen	Oxygen
Wood	49.65	6.23	0.92	43.20
Peat	55.44	6.28	1.72	36.56
Lignite	72.95	5.24	1.31	20.50
Bituminous coal	84.24	5.55	1.52	8.69
Anthracite	93.50	2.81	0.97	2.72

Table 12.4

(i) In what forms are carbon, hydrogen and oxygen lost as rank increases?
(ii) Why does peat have much more nitrogen than wood?
(iii) Why does coal of high rank have a higher calorific value than coal of low rank?
(c) The weight per cent of fixed carbon in peat and coals is approximately as follows.
Peat 28; lignite 38; bituminous coal 65; and anthracite 93.
(i) What is fixed carbon? Why is the percentage of fixed carbon less than the percentage of total carbon?
(ii) Assuming no loss of fixed carbon, how much lignite, bituminous coal and anthracite would be produced from 100 t of peat?

12.5 During electrical wireline logging, rock properties are measured by a logging tool or sonde which is slowly wound up a borehole.
(a) In constructing logs of radioactivity, gamma-ray emissions may be measured. Why do shales tend to exhibit higher levels of emission than sandstone, limestone, halite, coal and anhydrite?
(b) Neutrons emitted by the logging tool are absorbed by chemically-combined hydrogen in the rock or pore fluids.
(i) Why do porous rocks tend to absorb neutrons more efficiently than non-porous rocks?
(ii) Why does gypsum absorb neutrons more efficiently than anhydrite? Neither is porous.

(c) To produce an induction log, artificially-produced electrical currents in the rock induce a current in the receiver. The strength of the received signal is proportional to the electrical conductivity of the rock.
Why are some rocks better conductors of electricity than others? What rock types may be poor conductors?

(d) In sonic or acoustic logging, the velocities of sound waves emitted by the logging tool are measured. Why do sound wave velocities depend partly on the porosity of the rock?

(e) What is the value of having a large number of well logs in a potential oilfield?

12.6 Table 12.6 shows liquid petroleum reserves, production and consumption for various parts of the world for 1960 and 1980. Figures are millions of tonnes.

	Reserves		Production	
Area	1960	1980	1960	1980
Western Europe	236	3162	15	125
North America	4835	5000	412	563
Middle East	25 091	49 599	261	939
Africa	1110	7555	14	300
Latin America	3433	9519	195	300
Sino–Soviet area	4589	11 822	167	724
Far East and Australasia	1494	2689	27	134

Table 12.6

(a) (i) Why did every area show an increase in reserves even though petroleum was being produced between 1960 and 1980?

(ii) Explain why North America showed only a small increase in reserves while other areas showed large increases.

(iii) What were world reserves in 1960? What was the total production in 1960?
In 1960, in what year did it seem that liquid petroleum reserves would be exhausted?
What were world reserves in 1980? What was the total production in 1980?
In 1980, in what year did it seem that liquid petroleum reserves would be exhausted?

(b) (i) What is the limetime of a resource? How can the lifetime be extended?

(ii) What are renewable and non-renewable resources?

(iii) What is the place value of a resource? Name two resources of high place value and two of low place value.

(iv) In 1850, production of a resource material was 1 million tonnes. Until 1900, production increased by 2% a year. From 1901 to 1950, production increased by 3% a year. Thereafter, production increased by 4% a year.

Draw a graph showing production in ten-yearly steps from 1850 to the year 2000.

Here, resource production shows exponential growth. What is exponential growth?

How long does it take for production to double from 1850; from 1900; and from 1950? In each case what was the annual production before and after doubling?

Why would doubling of demand after 1950 cause bigger difficulties for producers than any previous doubling?

12.7 The Orkney Islands have subdued topography with shallow lochs. They are underlain by flagstones, siltstones, sandstones and marls which have low intergranular permeability. Joints provide moderate secondary permeability. In places, dykes are present. The aquifers are unconfined but recharge is hindered by partial cover of boulder clay and peat. Annual precipitation is 875 mm and evapotranspiration is 450 mm. The rain has high levels of sodium chloride.

Table 12.7 gives information on boreholes. Some properties of sea water are also given.

Borehole	Yield by pumping (litres per second)	Eh (mV)	Main ions (mg l^{-1})				
			Ca^{2+}	Na^+	Cl^-	SO_4^{2-}	HCO_3^-
1	0.2	400	57.3	40.2	72.3	8.3	189
2	1.0	400	52.3	52.8	84.2	6.5	229
3	1.0	350	88.0	43.9	62.9	10.1	274
4	2.0	360	71.5	44.2	73.4	7.2	218
5	1.0	400	55.8	50.3	78.0	4.4	262
6	1.0	400	50.0	44.4	81.2	4.4	289
7	2.0	240	353.8	40.3	45.8	234.6	380
8	3.0	300	54.0	26.6	47.1	6.3	272
9	12.0	280	90.0	49.4	85.0	14.1	395
Sea water	—	—	400	10 500	19 000	2600	140

Table 12.7

Eh is the redox potential: positive values indicate oxidizing conditions and negative values are indicative of reducing conditions.

(a) How does an unconfined aquifer differ from a confined aquifer?
(b) What is evapotranspiration?
(c) Why does the rain in Orkney have high levels of sodium chloride?
(d) What effect do dykes have on groundwater movement?
(e) How does peat hinder recharge of the aquifers?

(f) On a triangular diagram plot SO_4^{2-}, Cl^- and HCO_3^- as percentages of the total of the three ions in the well and sea water.

(g) Has the water in borehole 7 been contaminated by sea water? If not, why does the water from borehole 7 differ from the water from the other boreholes?

(h) Why may Orkney surface water not be suitable for domestic use?

(i) What is the daily water yield from boreholes 1 and 9?

(j) What is drawdown?

(k) Under what conditions will rates of drawdown be fast or slow?

(l) What is a cone of depression?

(m) What is the significance of the low Eh values for the water from boreholes 7 and 9?

12.8 (a) Petroleum is thought to form from marine phytoplankton (plant plankton), as shown in Fig. 12.8.

Marine phytoplankton

▼

—————————— Sea floor ——————————

Partial decomposition by microorganisms. Carbon dioxide and methane given off. Sapropel formed.

▼

Shallow burial. Increase in temperature. Geochemical alteration. Loss of carbon dioxide, water, nitrogen and sulphur. Source rock rich in organic matter (kerogen). Source bed usually contains about 2% carbon.

▼

Further burial. Further geochemical alteration beginning at about 75 °C. Large molecules broken down. After a long time, crude oil and natural gas are formed. The oil and gas migrate to reservoir rocks.

Fig. 12.8

(i) What is sapropel? Why do sapropels tend to form in restricted basins and on continental slopes?
(ii) What is kerogen?
(iii) Why do source beds tend to form in areas where deposition is neither too fast nor too slow?
(iv) Dry marine phytoplankton and crude oil have the following compositions.

	Elements (weight per cent)				
	Carbon	Hydrogen	Oxygen	Sulphur	Nitrogen
Marine phytoplankton	49.9	6.9	37.8	0.3	5.1
Crude oil	87.0	11.0	1.5	0.3	0.2

Table 12.8A

What percentage changes in the quantities of the elements are brought about by oil formation?
(v) Explain these apparent anomalies:
1 Sulphur shows no percentage alteration yet it is lost by geochemical alteration.
2 Carbon and hydrogen show a percentage increase yet both are lost by biochemical and geochemical alteration.
(vi) A chemical change which accompanies the formation of crude oil is the alteration of carbohydrate to carbon dioxide and to other chemicals of a particular type. Of what type are these other chemicals likely to be?

(b) (i) Crude oil is thought to form during burial when the temperature reaches 75 °C, while gas is thought to form when the temperature reaches 170 °C. Complete Table 12.8B to show the depths at which these temperatures are reached under the geothermal gradients shown.

Geothermal gradient ($°C\,km^{-1}$)	Depth at which temperature of 75 °C reached (km)	Depth at which temperature of 170 °C reached (km)
10		
20		
30		
40		
50		
60		

Table 12.8B

(ii) Draw a graph to show the range of depths within which crude oil forms under different geothermal gradients.
(iii) Within what depth range will oil form under a geothermal gradient of 25 °C km^{-1}?

(iv) What geothermal gradient will initiate oil formation at a depth of 5 km? What geothermal gradient would initiate gas formation at the same depth?

(c) Table 12.8C shows how the time for which a source rock has been buried affects oil and gas formation.

Time for which source rock buried (Ma)	Temperature at which oil begins to form (°C)	Temperature at which gas begins to form (°C)	Temperature at which oil destroyed and gas lost (°C)
10	75	170	410
50	58	135	300
100	52	120	270
200	45	105	240
300	40	100	230
400	35	95	220
500	33	92	210
600	30	89	200

Table 12.8C

(i) Draw a graph of the temperatures at which oil and gas begin to form and oil is destroyed and gas lost, against the time of burial of the source rock.

(ii) A source rock 150 Ma old has produced oil but no gas. Within what range of temperature has it been buried?

(iii) A source rock buried for 40 Ma has reached a temperature of 350 °C. State the form of any petroleum contained within this rock.

12.9 In the United Kingdom, sources of geothermal energy are hot ground water in sedimentary rock and hot dry rocks such as granites.

(a) Fig. 12.9A shows the main Palaeozoic and Mesozoic sedimentary basins and the positions of granite batholiths. Fig. 12.9B shows surface heat flow in millijoules per square metre per second ($mJ\,m^{-2}\,s^{-1}$).

(i) With what types of geological feature are high values of heat flow associated? Why are high values associated with these features?

(ii) Why are Palaeozoic sedimentary basins less suitable geothermal sources than Mesozoic basins?

(iii) The granites produce about 4.7 microjoules per cubic metre per second ($\mu J\,m^{-3}\,s^{-1}$). How much heat would be produced annually by a granite intrusion with a volume of $10^3\ km^3$? How does your answer compare with the annual United Kingdom use (10^{18} J) of electrical energy?

(b) (i) What is a geothermal gradient?

(ii) What is the thermal conductivity of a rock?

(iii) The average geothermal gradient in the United Kingdom is about 20 °C km^{-1}. The Cornish granites are thought to have a temperature of 200 °C at a depth of 5.4 km

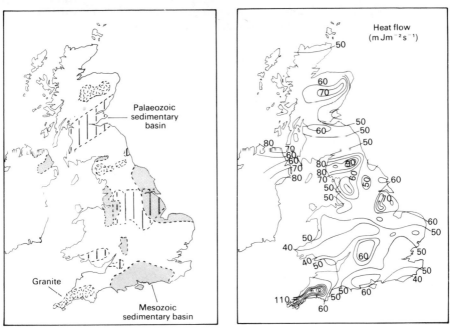

Fig. 12.9A   Fig. 12.9B

At what depth would a temperature of 200 °C be reached under the average United Kingdom geothermal gradient? What is the geothermal gradient associated with the Cornish granites?
(In both cases assume ground surface temperature to be 10 °C.)

(iv) How will a covering of rocks having high thermal conductivities affect the geothermal gradient associated with a granite batholith?

(v) If a granite is covered by rocks of low thermal conductivity, how will its potential as a source of geothermal energy be affected?

(vi) Why do thick sedimentary basins often have high geothermal gradients?

(c) How is heat extracted from granites and sedimentary rocks? For what purposes can the steam or hot water be used?

12.10 A hillside or embankment of granular material may fail along a slip plane (Fig. 12.10A). The weight (W) of the material above the plane of failure may be resolved into a force (N) vertical to the slip plane and a shear force (T) parallel to the slip plane. The vertical force tends to hold the two sides of the slip plane together but the shear force tends to cause sliding.

Since stress = force/area, the vertical stress, σ (*Greek sigma*) = N/A, and the shear stress which causes failure τ, (*Greek tau*) = T/A, where A is the area of the slip plane.

The coefficient of friction, μ (*Greek mu*) = $T/N = \tau/\sigma = \tan\phi$, where ϕ (*Greek phi*), the angle of friction, is the angle between W and N.

The shear strength of the material is equal to the shear stress which causes failure:

$$\text{shear strength} = \tau = \mu\sigma = \sigma\tan\phi$$

Many granular materials (e.g. moist clay) are cohesive or sticky. Cohesion is the force over and above internal friction which tends to hold the material together. For a cohesive material:

$$\text{shear strength} = c + \sigma \tan \phi$$

where c is cohesion.

When the material is waterlogged, the water provides a lifting force at right angles to the slip plane so the stress normal to the slip plane is reduced. In consequence, shear strength is reduced.

(a) (i) For a sand and gravel mixture $\phi = 40°$ and the normal stress at failure is 180 kN m^{-2}. The mixture has no cohesion. What is the shear strength of the material?

(ii) For a soil $\phi = 21°$ and $c = 170 \text{ kN m}^{-2}$. The normal stress at failure is 260 kN m^{-2}. What is the shear strength of the soil?

(b) Table 12.10A shows how properties of London Clay in cuttings change with time.

Age of cutting (years)	Cohesion $(kN\,m^{-2})$	ϕ (degrees)
0	15.3	20
25	6.2	17.5
40	3.8	17
60	0.0	16

Table 12.10A

(i) With the passage of time, what happens to the shear strength of the Clay? Explain your answer.

(ii) How will the change in shear strength make itself evident in road and railway cuttings?

(iii) Motorway cuttings in the Clay often slope at angles in excess of 18°. What would be a more appropriate angle?

(c) (i) As the steepness of a slip plane increases, what happens to the values of shear and normal stress? Would slip be more or less likely on a steeply-inclined plane?

(ii) Does water pressure in a saturated granular material produce shear stress within the material? Explain your answer.

(iii) In what ways does rainfall affect the stability of a slope?

(d) When a granular material is tipped or dumped it tends to form a conical heap. The angle between the surface of the cone and the horizontal is the angle of repose. Explain why the pairs of materials shown in Table 12.10B have different angles of repose.

(e) To find how compacted and loose sand behave when they are subjected to shear stress, they were held under pressure then pushed sideways.

(i) What mechanisms are responsible for the different types of behaviour shown by the compacted and loose sand in Fig. 12.10B?

(ii) Why do the shear strengths of both samples eventually become equal?

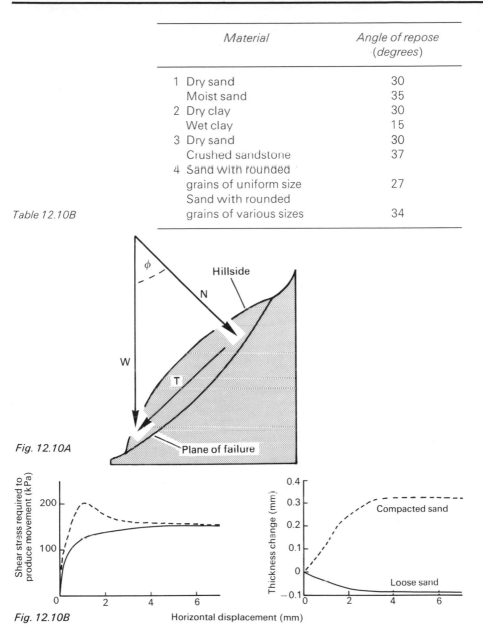

Material	Angle of repose (degrees)
1 Dry sand	30
Moist sand	35
2 Dry clay	30
Wet clay	15
3 Dry sand	30
Crushed sandstone	37
4 Sand with rounded grains of uniform size	27
Sand with rounded grains of various sizes	34

Table 12.10B

Fig. 12.10A

Fig. 12.10B

(f) Road construction in steep terrain is a major cause of landslides. Material dug from the hillside is called *cut* while material used to smooth-out hollows or to form embankments is called *fill*.
　(i) How will cutting and filling affect slope stability?
　(ii) How can slope disturbance be minimized?
　(iii) Why should embankments be thoroughly compacted? Why should large pieces of plant material (e.g. tree roots) be removed from the fill?

(iv) How would slope stability be affected by blasting?

(v) Sometimes, cut material cannot be used as fill. Where should such waste be disposed of to cause minimum disturbance to the slopes?

(vi) In balanced road construction, all of the cut material is used as fill. What is the advantage of this type of construction?

(vii) Types of road construction in steep terrain are shown in Fig. 12.10C. Which type of construction would be most suitable? Explain your answer.

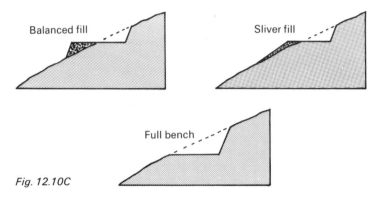

Balanced fill

Sliver fill

Full bench

Fig. 12.10C

(viii) Inadequate drainage of roads is a major cause of failures in steep terrain. How can drainage be improved? Why should natural drainage be impeded as little as possible?

(ix) Why should trees be left on slopes above roads?

(x) In areas where potential rotational failures exist, where on the slope should the road be sited?

(xi) A road 4 m wide is constructed in a hilly area. The embankments have a slope gradient of 1.5 and the cut slopes have a gradient of 0.5. Table 12.10C shows the disturbance created on slopes of different gradient. Draw a graph of hillslope gradient against width of disturbed area. What general relationship is shown by the graph?

Hillslope gradient	Width of disturbed area in planar projection (m)
0.2	5.0
0.3	5.5
0.4	6.5
0.52	10.0
0.6	15.0
0.63	20.0
0.66	30.0

Table 12.10C

12.11 (a) Fig. 12.11 shows a proposed dam site in a narrow valley. From the diagram, select four features which would make the site unsuitable. Explain your answer.

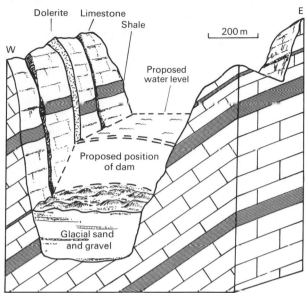

Fig. 12.11

(b) Investigation of another potential dam site showed it to have these properties:

The rocks are folded into a syncline which plunges upstream. There are no faults but joints are common. The river carries 10 tonnes of sediment per day in winter and 100 tonnes per day in summer. The river comes from mountains with glaciers. Water requirement is greatest in summer.

Give two reasons why the site may be suitable and two reasons why the site may be unsuitable.

12.12 A reservoir was found to be leaking at the rate of 1 million litres per day. When the water level was lowered for dam site investigation, leakage was reduced to 0.7 million litres a day. A pit dug at position P in Fig. 12.12 found badly-weathered sandstone with fissures running towards the dam. Fluorescein dye added to the water in the pit appeared in water below the dam 20 minutes later.

Drilling along the crest of the dam showed no cavities in the clay core of the dam. On drilling into the rock under the dam, there was sudden loss of drilling fluid, sudden drop of the drilling rods and poor core recovery. To test the permeability of the rocks, water was injected under pressure and the rate of water flow from the end of the drill was measured (Table 12.12A).

In an effort to prevent leakage from the reservoir, cement grout was injected at a pressure of 15 bars into the rocks under the dam. The grout was injected in mixes of increasing strength (Table 12.12B). To begin with, grout was injected at 12 m intervals. Grout was then injected half way between the original holes. Finally, grout was injected half-way between the previous holes. The amount of grout injected at each stage is shown in Table 12.12C. The grout curtain reduced leakage to an acceptable level of 2000 litres per day.

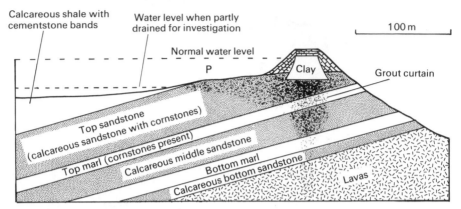

Fig. 12.12

Rock	Depth from crest of dam (m)	Flow of water from drill (litres per minute at 10 bars)
Top sandstone	8.4–13.6	618.0
Top marl	13.6–16.5	13.0
Middle sandstone	16.5–24.1	166.2
Bottom marl	24.1–27.3	2.1
Bottom sandstone	27.3–30.6	1.1
Lava	30.6–41.4	1.8

Table 12.12A Leakage tests from drill

Mix	Water/cement ratio	Tanks used in succession
1	10/1	4
2	5/1	5
3	8/3	10
4	8/6	10

Table 12.12B Grout mixes

Rock	Tonnes of grout injected per metre			Total tonnes of grout injected
	Stage 1 injection	Stage 2 injection	Stage 3 injection	
Top sandstone and top marl	7.13	5.2	—	56.29
Middle sandstone	1.8	0.48	0.06	19.3
Bottom marl, bottom sandstone and lavas	1.2	0.17	—	0.69

Table 12.12C Quantities of grout injected

(a) In what way do marl, cornstone, cementstone and calcareous sandstone and shale provide unsuitable reservoir foundations?
(b) How can you tell that the topmost shale is permeable?
(c) When drilling into the rock under the dam, why was there loss of drilling fluid, drop of drilling rods and poor core recovery?
(d) When water was injected, what was the leakage per metre from each rock unit?
(e) Why is the top sandstone the most permeable unit?
(f) Despite being beneath the top marl, the middle sandstone is the more permeable. Why does this difference occur?
(g) By means of a sketch, show the direction of water flow under the dam.
(h) Why was the strength of the grout mix progressively increased?
(i) Why was the number of tanks used in succession increased for the first three grout mixes?
(j) Why did the tonnes of grout injected decrease progressively from the Stage 1 to the Stage 3 injections?

12.13 What are the functions of the roadside engineering features shown in the photographs?

(a)

(b)

12.14 (a) Environmental geology is the study of geological factors relevant to our health, safety and welfare. One aspect of environmental geology is the study of ground stability and its effects on buildings.

    (i) In old mines, pillars between worked stalls held up the roof. How do such workings cause differential subsidence?

    (ii) How does it come about that houses built on concealed, backfilled quarries suffer structural damage?

    (iii) How would you recognize the effects of subsidence on buildings?

    (iv) Why are buildings on drift deposits liable to suffer the effects of subsidence?

    (v) How can the extraction of ground water cause subsidence?

    (vi) How may extraction of soil water by tree roots cause structural damage to a house? How may felling the tree cause further damage?

    (vii) If cavernous limestones are dewatered by mining activity, what may happen to overlying sediments?

(b) Subsidence from longwall coal workings takes place as shown in Fig. 12.14A.

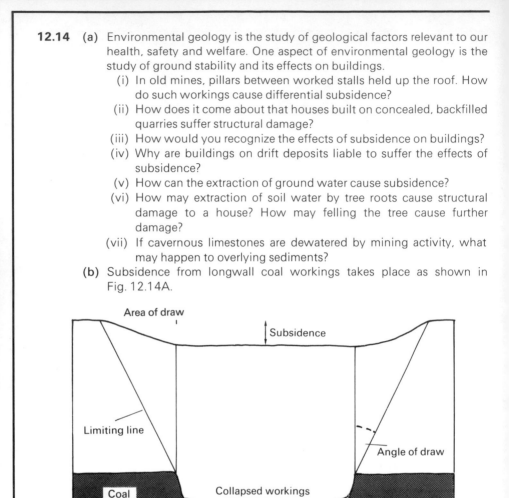

Fig. 12.14A

    (i) In which part of the area of draw are buildings and roads under tension? What would be the effects of tension? In what part of the area of draw are buildings and roads under compression? What would be the effects of compression?

    (ii) How does it sometimes happen that subsidence is concentrated along narrow zones?

    (iii) What happens to the area of draw as the depth to the coal seam increases?

    (iv) If the coal face advances rapidly, what type of movements will occur at the surface? What will happen if the face is abandoned?

    (v) How can old mine workings be detected?

(c) The amount of subsidence produced by longwall workings depends on the thickness of the coal seam, the depth to the seam and the width of the workings. Fig. 12.14B shows the maximum subsidence as a fraction of seam thickness for workings of various widths at various depths.

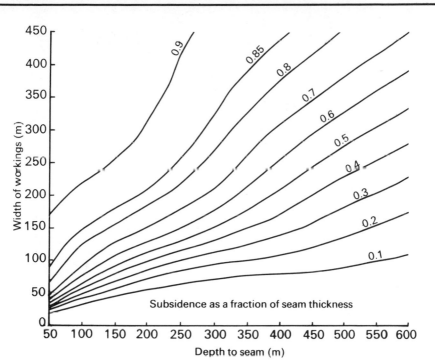

Fig. 12.14B

(i) How much subsidence could be produced by the following workings?

	Seam thickness (m)	Width of workings (m)	Depth to seam (m)
A	1.5	300	300
B	1.5	300	600
C	1.5	450	300
D	3.0	300	300

Table 12.14

(ii) What general conclusions can be drawn from these results?

(d) Fig. 12.14C overleaf shows the relationships between oil production, subsidence and water injection in the Wilmington Oilfield, Long Beach, California.

(i) What caused the subsidence? Why did the land surface recover when water was injected?

(ii) What particular problems would be caused by subsidence in coastal districts?

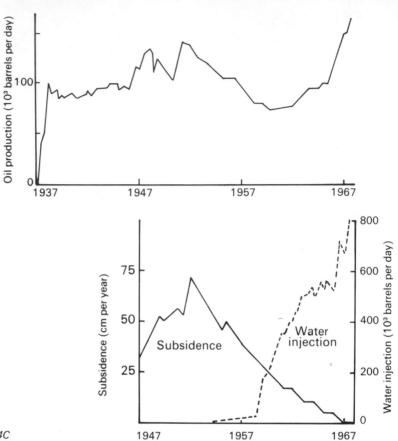

Fig. 12.14C

12.15 In the body, iodine is used to make the essential hormone thyroxine. Deficiency of iodine in the diet causes goitre which shows itself as a swelling of the thyroid gland. In children severe deficiency causes cretinism. Table 12.15A shows the amounts of iodine in various materials.

Material	Amount of iodine (parts per million)
Igneous rocks	0.5
Soils derived from igneous rocks	9.3
Sedimentary rocks	1.5
Soils derived from sedimentary rocks	3.8
Air	0.0005
Rainwater	0.0006
Seawater	0.06

Table 12.15A

Table 12.15B shows the relationship between the occurrence of goitre in children and the amount of iodine in soil, water and locally-produced foods for towns in the Republic of Ireland.

County in which town situated	Percentage of children with goitre	Amount of iodine (parts per million)				
		Soil	Water	Milk	Bread	Potatoes
Tipperary	65	30.71	0.005	0.015	0.047	0.006
Leix (Laois)	40	30.10	0.017	0.035	0.126	0.117
Mayo	10	50.50	0.004	0.036	0.107	0.070
Galway	0	143.90	0.201	0.556	0.183	0.056

Table 12.15B

Mayo and Galway are on the west coast; Tipperary and Leix are both inland. Galway has the greatest exposures of igneous rocks.

(a) How does iodine get into air and rainwater?

(b) A soil contains 140 parts per million (ppm) of iodine. Assuming that all of the iodine from an igneous rock remains in the soil, how much igneous rock would have to be chemically weathered to produce a tonne of soil? If the rock had a density of $3.0\,t\,m^{-3}$, what thickness of rock of section 1 m^2 would have to be weathered-down? If iodine is lost from soil, how would these figures be affected?

(c) Is it likely that all of the iodine in Galway soils has come from weathered rock? Explain your answer. How else may iodine have entered the soil?

(d) By how many times is iodine in the soil above igneous rock concentrated, relative to the igneous rock? By how many times is iodine in the soil above sedimentary rock concentrated, relative to the rock? The reason for the difference has not been fully explained. Nevertheless, can you suggest any way in which the differential concentration may have come about?

(e) Explain why Galway is the only area free of goitre.

(f) Why is goitre particularly common in recently glaciated areas such as the Alps, Pyrenees, Himalayas and Andes?

12.16 Remote sensing is the use of electromagnetic radiation (light, infra-red and radar) to gather Earth surface data from aircraft and satellites. Rock identification depends to some extent on spectral reflectance. In photographs, variable reflectance appears as different shades of grey (tone) or as colour variation. Reflectance depends on the electromagnetic wavelength and on the composition, structure and surface properties of the rock. Variations in iron, water, carbonate and clay contents are largely responsible for differences in spectral reflectance. Table 12.16 overleaf shows the wavelengths at which high levels of absorption occur.

Substance in rock	Wavelengths strongly absorbed (micrometres, μm)			
H_2O, OH^-		1.4	1.9	2.2
Fe^{2+}	1.0		2.0	
Fe^{3+}	0.7 0.85 (from stain on sand grains)			
CO_3^{2-}		1.9		2.3

Table 12.16

Note: The presence of carbonaceous materials may obscure spectral features.

(a) Fig. 12.16A shows laboratory reflectance spectra of various rocks. Account for the differences between the spectra of the following pairs of rocks: granite and olivine peridotite (dunite); fossiliferous limestone and red sandstone; serpentine marble and hornblende schist.

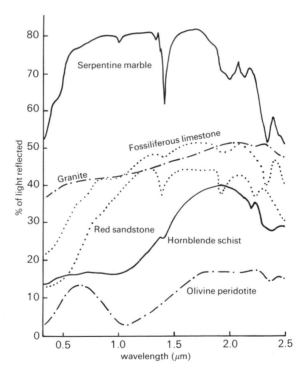

Fig. 12.16A

(b) Fig. 12.16B shows laboratory reflectance spectra of fresh and weathered andesite. Account for the differences in the spectra.
(c) What effects on tone would result from a single rock type being partly in the sun and partly in the shade?
(d) Moisture reduces reflectance in both the visible and infra-red regions. How would the same rock type appear if different moisture contents were present?

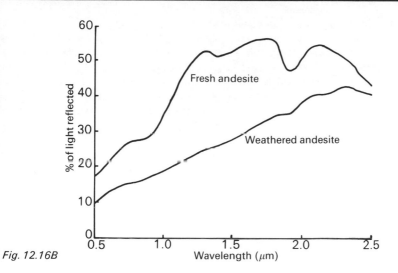

Fig. 12.16B

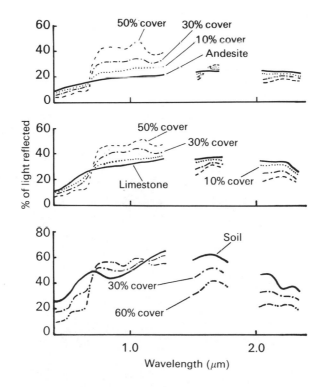

Fig. 12.16C

(e) Combining black and white film with a filter which removes all wavelengths shorter than 0.7 µm allows the production of photographs taken in reflected short infra-red radiation. The process eliminates the atmospheric scatter which occurs with visible light. What effect will this elimination have on the sharpness of the images? Since water absorbs infra-red radiation, how will rivers and lakes appear in such photos?

(f) At what time (day or night) would it be best to record images from emitted long infra-red radiation? Explain your answer.

(g) Give four features found in photographs which may change with time.

(h) In arid areas vegetation may be specific to certain rock types. In other regions, vegetation is non specific. What factors will influence the type and extent of vegetation?

(i) Images are composites which show the reflectance of rock, soil, vegetation and other materials. Fig. 12.16C shows the reflectance spectra of andesite, limestone and soil with various degrees of grass cover. What effects are produced by grass cover? At which wavelengths are the slopes of the spectral curves most strongly affected?

12.17 The satellite photograph shows part of Tennessee in the Appalachian Mountains, USA. The Sun is shining from the east–south-east.

(a) From the photograph, draw a sketch map showing the main geological features of the area. At times, you may find it useful to view the photograph from a near-horizontal position.

(b) Why have the rivers been dammed?

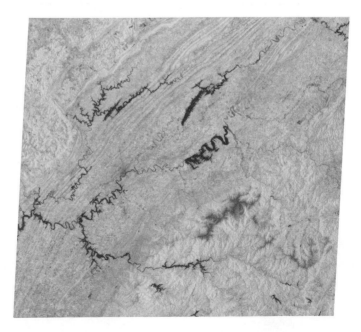

12.18 (a) At some time in your study of science you will come across the terms defined below. The definitions have come from well-known dictionaries.

Fact: Something that has really occurred or is the case.

Hypothesis: A supposition or conjecture put forth to account for known facts. A provisional supposition which accounts for known facts and serves as a starting point for further investigation, by which it may be proved or disproved.

Rule: A fact, or statement of one, that holds generally good; that which is normally the case.

Theory: A hypothesis that has been confirmed or established by observation or experiment and is propounded or accepted as accounting for the known facts. A statement of what are held to be the general laws, principles or causes of something known or observed.

Principle: A highly general or inclusive theorem or 'law' exemplified in a multitude of cases.

Law: A theoretical principle deduced from particular facts expressible by the statement that a particular phenomenon always occurs if certain conditions be present.

Which of the defined terms applies to the following statements?
  (1) In the fossil record, vertebrates come after invertebrates.
  (2) Vertebrates developed from invertebrates.
  (3) Equal volumes of gases at equal volumes and pressures contain equal numbers of molecules.
  (4) The Earth's core consists of iron and nickel with minor amounts of other substances.
  (5) Vertebrates evolved from invertebrates through the operation of natural selection.
  (6) The force of attraction between two objects is proportional to the product of the masses and inversely proportional to the square of the distance between the masses.
  (7) Continental drift takes place because the continents are carried on moving lithospheric plates.
  (8) Landscapes change through the operation of a cycle of erosion.
  (9) The temperature at the centre of the Earth is 5000 °C.
  (10) The continents are moving away from each other because the Earth is expanding.
  (11) The disappearance of fossil life-forms and the appearance of new ones in younger beds shows that life was continually being destroyed and recreated.
  (12) The dinosaurs became extinct because of the effects of an asteroid striking the Earth.
  (13) The last Ice Age began because increased volcanic activity produced atmospheric dust which blotted out sunlight.
  (14) During the Upper Cretaceous period, sea covered much of Britain.
  (15) Crystals of the same substance have the same interfacial angles between corresponding faces.
  (16) The Earth is 4600 million years old.

(17) Africa and South America were once joined.

(18) Diamond and graphite are different forms of carbon.

(19) Eyeless trilobites lived by burrowing in mud.

(20) The solar system formed from a nebula of dust and gas.

(21) In a series of coal seams, the fixed carbon increases and the volatile matter decreases with depth.

(22) If one stratum overlies another, the top stratum is younger than the bottom one.

(23) The external and internal processes we recognize today have been operating unchanged and at the same set of rates throughout most of the Earth's history.

(24) Where there is a gradation from relatively rich to relatively low-grade material in a mineral deposit, the tonnage of ore increases geometrically as the grade decreases arithmetically.

(25) Strata that contain fossils of the same species of animals and plants were produced in the same period.

(26) The low-density crust behaves as if it floats on the high-density mantle.

(27) If fragments of one rock are found in another rock, the enclosing rock must have been formed later than the rock material of the included fragments.

(28) Water-laid sediments are deposited in strata that are not far from horizontal, and parallel or nearly parallel to the surface on which they are accumulating.

(29) Among the different organisms which make up a population of a given species, individuals having advantageous characteristics contribute more offspring to the succeeding generation than those having disadvantageous characteristics. The composition of the population changes to include more better-adapted organisms.

(30) When waves pass from one medium to another, the ratio of the sine of the angle of incidence to the sine of the angle of refraction is a constant for any pair of media.

(b) (i) Also in your studies you will come across the word 'model'. Here are two ways in which the term has been defined.

    1 A model is a description, a collection of statistical data, or an analogy used to help visualize, often in a simplified way, something that cannot be directly observed (e.g. the structure of the atom; the interior of the Earth).

    2 Models are essentially statements about reality made with various levels of abstraction from reality. These statements may be made mechanically, mathematically or verbally. Models fall short of being theory, but models may develop into theory.

    Suggest three ways in which models may be used by scientists.

(ii) The Hubbert model for estimating reserves is based on two assumptions:

    1 The amount of reserves remaining at any time is the difference between the initial reserves and the total production to that date.

    and

    2 In a complete cycle of extraction, production rises exponentially to a maximum then declines exponentially to zero as the resource

material becomes scarcer and so more difficult to extract (Fig. 12.18A).

What factors in these assumptions may lead to inaccuracies in estimating reserves?

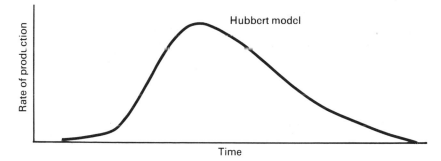

Fig. 12.18A

(c) North Sea oil production is very sensitive to changes in the price of oil because many fields are only marginally profitable. A price change of only 2 or 3 dollars a barrel makes the difference between some fields making a profit or a loss.

Fig. 12.18B (below and overleaf) shows models of future North Sea oil production. The models work by triggering off the development of new fields when a post-tax profit of at least 10% is achieved. (The models were developed by Professor A. G. Kemp, University of Aberdeen, 1987.)

Fig. 12.18B (i)

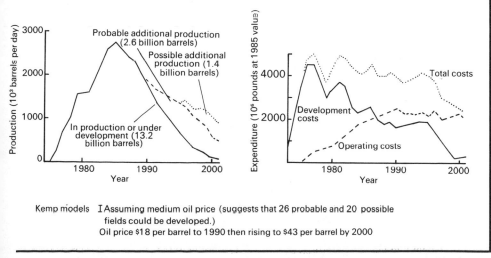

Kemp models I Assuming medium oil price (suggests that 26 probable and 20 possible fields could be developed.)
Oil price $18 per barrel to 1990 then rising to $43 per barrel by 2000

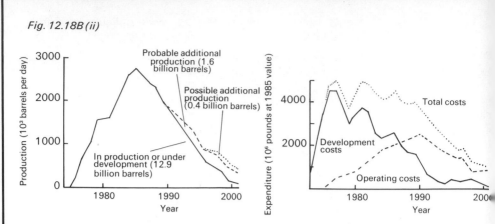

Fig. 12.18B (ii)

II Assuming low oil price (suggests that 17 probable and 6 possible fields could be developed)
Oil price $18 per barrel to 1990 then rising by 1% per year in real terms to 2000

   (i) On what assumptions are the models based? Answer in general terms only.
  (ii) Why do predicted production rates differ in the two models?
 (iii) How does Kemp's production cycle compare with that suggested by Hubbert?
 (iv) How would high or very high oil prices affect future production from the North Sea?
  (v) Why have development costs in the North Sea fallen from 1976?

PLANET EARTH

1.1 (a) Temperatures would have been highest in the middle of the nebula. Mostly silicates, Fe–Ni alloy and FeS.

(b) 1 Chemical composition. A planet with a high proportion of Fe–Ni alloy and silicates would be relatively dense. A planet consisting largely of hydrogen, helium, methane and ammonia would have a low density.

2 Size. Very high pressure inside large planets squeezes atoms closer together so increasing internal density.

(c) These terrestrial planets condensed from high-density material near the centre of the nebula. The other planets condensed mostly from low-density gaseous material in the cold outer parts of the nebula.

(d) Mercury has a very large core. Being nearest the Sun, it apparently condensed from that part of the nebula richest in Fe–Ni.

(e) When the early Sun began to shine, it shed a large quantity of material. This material formed a strong solar wind which swept gases to outer parts of the nebula.

(f) Both would increase.

(g) Heat.

(h) Carbon was present in the nebular material from which the planets formed. Fe would sink to form a core.

1.2 (a) Age relationships are found by studying photographs. The relative ages of lava flows, ejecta blankets and craters can be found from cross-cutting and overlying relationships. Rocks brought back by Apollo astronauts have been dated radiometrically.

(b) 1 Anorthosite crust;   2 most craters of terra;   3 mare lava flows;   4 crater Q;   5 crater R;   6 crater P;   7 volcano and sinuous rille;   8 fault.

(c) Older areas are more heavily cratered than younger areas.

(d) Crater rays are long, bright streaks which radiate from relatively young craters. They consist of material ejected from craters during meteorite impacts.

(e) Crater rims are gradually smoothed off by bombardment from micro-meteorites. Exposure to solar radiation causes rays to lose their brightness.

(f) Regolith consists of debris from bombardment by meteorites, cosmic rays and the solar wind. It contains glass beads produced by rock melting during meteorite impact. Regolith accumulates at the rate of about 1 mm per million years.

(g) In crater rims older material may overlie younger material. During impact, deeper, older material is last to be ejected.

(h) When the surface of a lava flow cools and solidifies, the lava inside may still flow. When flow ceases, a hollow lava tube or lava tunnel may remain.

(i) Crater chains may be produced by volcanic activity along the line of a fracture or by large fragments ejected in a volley from an impact crater.

1.3 **(a)** Land area: 148.9×10^6 km². Sea floor area: 333.3×10^6 km².

(b) Land:

Height (m)	% total area
0–1000	22.11
1000–2000	4.69
2000–3000	2.32
3000–4000	1.20
4000–5000	0.46
above 5000	0.10

Sea floor:

Depth (m)	% total area
0–1000	3.78
1000–2000	3.16
2000–3000	5.08
3000–4000	14.68
4000–5000	24.70
5000–6000	16.82
6000–7000	0.83
below 7000	0.00

(c) See Fig. A1.3.

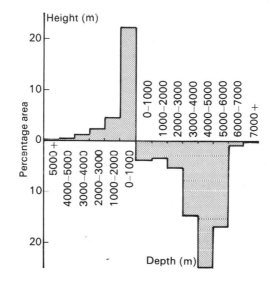

Fig. A1.3

(d) Land: 0–1000 m; sea floor: 4000–5000 m.
If continental and oceanic crusts were similar, an even distribution of height and depth on either side of sea level would be expected. Instead, the step down from continent to sea floor suggests a significant difference between the crustal types.

1.4 **(a)** 1 Island arc;   2 oceanic trench;   3 abyssal plain;   4 continental rise;   5 continental slope;   6 continental shelf;   7 submarine canyon;   8 guyot;   9 seamount;   10 submarine fan;   11 oceanic ridge;   12 axial rift valley;   13 atoll;   14 fracture zone.

(b) Craton: A large area of stable continental crust. It has not been subjected to deformation since the Precambrian or early Palaeozoic. Often equated with shield. A shield consists mostly of Precambrian metamorphic and igneous rocks. Locations: Baltic Shield; Canadian Shield; much of Greenland and S.E. India; parts of W. Australia, Africa and eastern S. America.

Platform: A stable shield area covered by weakly-deformed sedimentary rocks. Locations: Great Plains of N. America; Steppes; much of Siberia; much of the Sahara and Saudi Arabia; much of S. America east of the Andes.

(c) Orogenic belt: A linear region of mountain building characterized by deformed rocks which are often metamorphosed and intruded by plutonic rocks.
Young ranges: Alps; Himalayas; Andes; Rockies; etc.
Old ranges: Scottish Highlands; Urals; Appalachians; parts of China, Mongolia and south-east USSR.

MINERALS

2.1 **(a)** (i) In an electrovalent bond, an atom loses one or more electrons to another atom. The atom losing electrons becomes a positively-charged cation; the atom gaining electrons becomes a negatively-charged anion. In a covalent bond, atoms share electrons. In both bonds atoms achieve stable electron configurations.

(ii) The number of immediately-surrounding ions to which it is linked.

(iii) Si^{4+}/O^{2-}: ratio of ionic radii $= 0.30$. Expected coordination number $= 4$.
Al^{3+}/O^{2-}: ratio of radii $= 0.36$. Expected coordination number $= 4$.
Mg^{2+}/O^{2-}: ratio of radii $= 0.47$. Expected coordination number $= 6$.

(iv) 1 High temperatures and low pressures tend to produce low density crystals with widely spaced atoms. The low coordination numbers reflect the loose packing of the atoms. Low temperatures and high pressures tend to produce dense crystals. The high coordination numbers reflect the tight packing of the atoms.

2 In Table 2.1B, the ionic radius ratio which would tend to cause a switch from 4- to 6-fold coordination is 0.41. The Al^{3+}/O^{2-} ratio (0.36) is sufficiently close to 0.41 to allow Al^{3+} to adopt 4- or 6-fold coordination depending on conditions.

(b) (i) Substances of differing compositions which have the same atomic structure and crystal form.

(ii) Solid solutions (also called mixed crystals) are solid, homogeneous mixtures of two or more substances. They are the result of ionic substitution rather than true mixing. Not all isomorphous substances form solid solutions.

(iii) Different crystalline forms of the same substance. Diamond and graphite are polymorphs of carbon.

(iv) Phases are substances which differ physically from each other; e.g., ice, water and steam are different phases of H_2O. A phase change is when one phase changes to another; e.g., with increased pressure, graphite changes to diamond. A phase diagram shows the stability fields of phases under conditions of changing pressure, temperature and concentration. One.

(v) When substances are mixed, the melting point of the mixture tends to be lower than the melting points of the pure substances. The eutectic mixture is the mixture with the lowest possible melting point.

(vi) 1 Because Mg^{2+} and Fe^{2+} have the same charges and similar sizes, they can readily substitute for each other. Anions and cations of the same size and in the same numbers tend to form crystals of the same type.

2 When Al^{3+} substitutes for Si^{4+} the feldspar loses a positive charge. To redress the balance of charges, Ca^{2+} (radius 99 pm) substitutes for Na^+ (radius 97 pm).

2.2 (a) (i) Relative density is the ratio of the mass of a given volume of a substance to the mass of an equal volume of water at a temperature of 4 °C. Also defined as the ratio of the density of a substance to the density of water at 4 °C.

(ii) Any impurity would almost certainly have a relative density different from that of the mineral. The relative density found by experiment would be intermediate in value between that of the mineral and the impurity.

(b) Forsterite (magnesium olivine, Mg_2SiO_4) has a relative density of 3.22; fayalite (iron olivine, Fe_2SiO_4) has a relative density of 4.39. Olivines of intermediate compositions have intermediate densities because Fe^{2+} (mass number 56) and Mg^{2+} (mass number 24) readily substitute for each other.

2.3 (a) Cleavage: The mineral breaks along atomic planes of weakness which are parallel to possible crystal faces. Fracture: The break does not follow a predetermined plane of atomic weakness. Common in minerals (e.g. quartz) where chemical bonds are of roughly equal strength. Fracture surfaces are often irregular.
1 Fluorite; 2 pyroxene; 3 amphibole; 4 calcite or dolomite; 5 halite or galena.

(b) Twinned crystal: A single crystal formed by the intergrowth of two or more crystals. One part of the twin is in reversed orientation relative to the next. Twins form when the crystal lattice changes its direction of growth.
1 Orthoclase; 2 staurolite; 3 rutile; 4 plagioclase; 5 fluorite; 6 microcline; 7 leucite; 8 gypsum.

(c) 1 Malachite; 2 sphalerite; 3 haematite; 4 apatite; 5 dolomite; 6 serpentine; 7 chalcopyrite; 8 tourmaline; 9 barite; 10 graphite.

2.4 (a) Tetrahedra are separate. The minerals are dense, hard and they have no well-defined planes of weakness. Other ions are tightly packed around the tetrahedra. Relative density (d) 3.2–4.4; hardness (H) 6–$7\frac{1}{2}$.

(b) Tetrahedra are in single chains. Each tetrahedron shares two corners with adjacent tetrahedra. Pyroxenes are dense and hard. The ions are quite tightly packed. The chains are tightly held together by cations. However, since the bonds between the chains are weaker than the bonds within the chains, cleavage runs between the chains in two directions which are nearly at right angles to each other. $d = 3.2–3.9$; $H = 5–6$.

(c) The tetrahedra are in double chains. Tetrahedra share alternately two and three oxygens. Amphiboles are less compact and less dense than pyroxenes. The double chains are not strongly held together by cations so distinct cleavage runs between the chains in two directions at about 60° to each other. Crystals tend to be long and fibrous. $d = 2.9$; $H = 5–6$.

(d) The tetrahedra are in sheets. Tetrahedra share three corners. Talc, mica and clay are not compact, hard or dense. The sheets are weakly held together so perfect cleavage runs between the sheets. The sheets form hexagonal planar networks which make the crystals look almost hexagonal. $d = 2.6–3.3$; $H = 1–3\frac{1}{2}$.

(e) The tetrahedra form a completely interlocking framework. All corners of the tetrahedra are shared. Strong bonds in three dimensions produce a rigid structure. The framework is open so quartz and feldspar have low densities. Quartz has no cleavage. Feldspar has good cleavage because of weaknesses introduced by ionic substitution. $d = 2.55–2.76$; $H = 6–7$.

2.5 **(a)** (i) Anorthite: 1550 °C; diopside: 1390 °C.

(ii) 1330 °C; anorthite.

(iii) Composition: anorthite 42%–diopside 58%. Temperature: 1270 °C. The temperature remains constant while the diopside and remaining anorthite crystallize. Further cooling takes place when crystallization is complete.

(iv) 1 This is a eutectic mixture. The melt would cool to point E then anorthite and diopside would crystallize together while the temperature remained constant.

2 Diopside begins to crystallize first. The melt cools along the diopside liquidus as diopside crystallizes. When the melt reaches point E, the anorthite and remaining diopside crystallize together.

(b) (i) Crystallization begins at 1450 °C; ends at 1290 °C.

(ii) Anorthite 83%–albite 17%.

(iii) Anorthite 50%–albite 50%.

(iv) Anorthite 15%–albite 85%.

(v) If the melt is viscous or if cooling is rapid, the solid plagioclase does not have time to react fully with the liquid. The first formed plagioclase has a composition of about anorthite 85%–albite 15%. Since reaction with the liquid is incomplete, the Ca-rich solid persists. With the removal of anorthite, the liquid becomes over enriched in albite and the crystal develops concentric zones which become progressively richer in albite. The outer zone of the crystal may contain very little anorthite. The overall composition of the zoned crystal is the same as that of the original melt (anorthite 60%–albite 40%).

2.6 **(a)** (i) A crystal is a body bounded by symmetrically-arranged plane faces. The external form is an expression of the regular internal atomic arrangement.

(ii) Grains are particles or crystals which make up rocks or loose sediments.

(iii) The regular spatial arrangement or network of atoms in a crystal.

(b) (i) Plane of symmetry (mirror plane): A plane which divides a crystal into two equal parts which are mirror images of each other.

Axis of symmetry (rotation axis): If a crystal is rotated about a line or axis, it may present the same view two or more times. An axis which presents the same view twice in a rotation is a diad; an axis which makes the crystal look the same three times in a rotation is a triad axis. A tetrad axis is an axis of 4-fold symmetry and a hexad axis is an axis of 6-fold symmetry.

Centre of symmetry: When every point on a crystal (e.g. a corner) can be joined to a similar point on the other side of a crystal by a line passing through the centre of the crystal, the crystal is said to have a centre of symmetry. Similar, parallel faces lie on opposite sides of the centre of the crystal. The similar faces are not mirror images of each other.

(ii) Crystals are made up of repeated, brick-like atomic blocks called unit cells. The unit cells which can be built together to completely fill a space cannot be of any shape. Building blocks with 5-fold symmetry cannot fill space and for this reason no crystals have this type of symmetry.

(c) (i) Habit: A description of the characteristic shapes of crystals arising from variation in the number, size and shape of faces.

Form: A set of crystal faces of the same type.

(ii) 1 Dog-tooth spar. Sharp pyramid (scalenohedron) and prism.
2 Nail-head spar. Blunt pyramid (flat rhombohedron) and prism.
3 Prismatic habit. Basal pinacoid (flat end faces) and prism.
(iii) Trigonal. Three vertical planes of symmetry; 1 triad; 3 diads; a centre of symmetry.
(iv) Cubic: 4 triads; tetragonal: 1 tetrad; orthorhombic: 3 diads; hexagonal: 1 hexad; monoclinic: 1 diad; triclinic: possesses a centre of symmetry only.

2.7 (a) (i) When a ray of light travels at an angle from one transparent substance to another, it is bent or refracted. Refraction occurs because the light travels at different velocities in the different materials. When a beam of light moves from air into a mineral, the light is slowed down and the beam is bent towards the normal.

(ii) The refractive index of a transparent material is the ratio of the velocity of the incident ray in a vacuum (or, approximately, in air) to the velocity of the refracted ray in the material. From Snell's Law:

$$\text{refractive index} = \frac{\sin i}{\sin r}$$

where i is the angle of incidence and r is the angle of refraction.

(iii) From Snell's Law:

$$\text{refractive index} = \frac{\sin 30°}{\sin 19°}$$

$$= 1.536$$

(iv) Mineral 1: The RI of the mineral is very close to that of Canada balsam. The mineral would merge with its background and be difficult to see. That is, it would have very low relief. The relief is described as being positive because the RI of the mineral is greater than the RI of the balsam.

Mineral 2: The RI of the mineral is distinctly different from that of the balsam. Since the mineral would stand out quite well from its background, it is said to have moderate relief. The relief is negative because the RI of the mineral is less than the RI of the balsam.

Mineral 3: The RI of the mineral is very different from the RI of the balsam. The mineral would stand out strongly from its background. It has high relief. The relief is positive because the RI of the mineral is greater than the RI of the Canada balsam.

(v) The calcite is showing double refraction or birefringence because it is forming two refracted rays from one incident ray. Since the two rays travel through the calcite at different velocities in different directions, the calcite has a different refractive index for each ray.

(vi) An isotropic mineral has the same internal properties in any direction through the crystal. Cubic minerals are isotropic because their atoms are arranged in very regular groups. Light rays travel at the same speed in all directions through the crystal so the mineral has only one value of refractive index.

An anisotropic mineral has different internal properties in different directions through the crystal. Light rays travel at different speeds in different directions through the crystal so the mineral has more than one refractive index. Non-cubic minerals are anisotropic.

(vii) Polarization colours are interference colours. When the beam of plane polarized light from the lower polar (the polarizer) enters an anisotropic mineral, the light is broken into two beams which vibrate in planes at right angles to each other. One beam travels faster than the other so its waves reach the upper polar (the analyser) slightly ahead of the waves in the other beam. Since the components of the two beams which pass through the upper polar are out of phase, the waves can cancel or reinforce each other by the process of interference. In this way, some wavelengths or colours are removed while others pass through.

(viii) For grains of constant thickness interference colours depend on the orientation of the grains and on the amount of double refraction. Grains which remain black throughout rotation have been cut at right angles to the length of the crystal. Such grains show no double refraction so they behave as if they were isotropic. For this reason no interference colours are produced.

Grains cut parallel to the length of the crystals have maximum birefringence so they produce the highest possible colours for the mineral on Newton's scale. The colours are brightest when the two refracted beams inside the crystal vibrate in planes at 45° to the vibration directions of the polars. Four times in a rotation of the stage the vibration planes in the crystal come to lie parallel to the vibration planes of the polars. When this happens, the grain goes black (goes to extinction). Light which passes through the lower polar passes through the mineral in the same plane. This light meets the upper polar at right angles to its vibration direction so it cannot pass through. See Fig. A2.7.

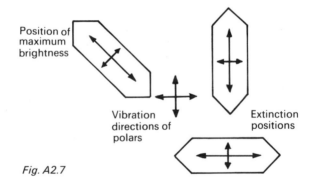

Position of maximum brightness

Vibration directions of polars

Extinction positions

Fig. A2.7

(b) 1 Hornblende;   2 chlorite;   3 biotite;   4 leucite;   5 plagioclase;
    6 calcite;   7 augite;   8 muscovite;   9 olivine;   10 orthoclase;
    11 quartz;   12 analcite;   13 andalusite.

IGNEOUS ROCKS

3.1 (a) (i) Essential minerals are those used in naming and classifying igneous rocks. An essential mineral is not necessarily a major constituent of the rock. An igneous rock may contain other, accessory minerals, which are not used in naming or classifying the rock.

(ii) Dolerite: plagioclase and augite.
Rhyolite: quartz and alkali feldspar (orthoclase, perthite, microcline, albite).

(b) (i) Intermediate. Diorite, microdiorite, andesite, syenite, microsyenite, trachyte.

(ii) Peridotite, dunite, kimberlite.

3.2 (a) (i) The inner mineral has been reacting with the magma to produce the outer mineral when crystallization ceased. The outer mineral is sometimes said to form a reaction rim or a reaction corona.

(ii) Early-formed olivine has not had time to react in the rhyolite because the viscous magma has been rapidly cooled. Slow cooling in the granite has allowed the olivine to react completely to form biotite and hornblende.

(b) If minerals such as olivine, pyroxene and Ca plagioclase are removed from the magma before they have time to react, the remaining magma will tend to assume an intermediate composition because it has been enriched in oxides such as SiO_2, H_2O, K_2O and Na_2O. Further fractionation may produce a watery magma rich in orthoclase, muscovite and quartz.

3.3 (a) Ophitic texture. Formed by the simultaneous crystallization of plagioclase and augite. The plagioclase produced numerous crystals by growing slowly from a large number of crystal nuclei. The large augite has grown rapidly to enclose the feldspars.

(b) Porphyritic texture. The large crystals (phenocrysts) have grown while the magma cooled slowly at depth. The magma has then been extruded or intruded at shallow depth and rapid cooling has produced the fine-grained groundmass.

(c) Graphic texture. On cooling, a eutectic mixture of quartz and orthoclase has produced an intergrowth as the minerals crystallized simultaneously.

(d) Spherulitic texture. May be produced by the devitrification of glass. May also be produced at moderate levels of supercooling in viscous magma when few crystal nuclei form but the potential growth rate of crystals is still high.

(e) Trachytic texture or flow structure. Flow of a viscous lava aligns the feldspar crystals.

(f) Banded texture. Formed by the settling of high-density cumulus crystals to the floor of a magma chamber.

3.4 (a) Q: A 40%, B 30%, C 30%; R: A 60%, B 10%, C 30%.

(b) U: granite; V: granodiorite; W: peridotite; X: syenite; Y: diorite; Z: gabbro.

3.5 (a) The partial melt contains much less MgO than the original rock. The other oxides increase to various extents in the partial melt. (For example, SiO_2 increases by a factor of 1.1; Al_2O_3 increases 5.1 times; and Na_2O increases 10.8 times.)

(b) (i) Relative formula mass of Mg_2SiO_4 ($= 2MgO.SiO_2$) $= 140.7$
Relative formula mass of $SiO_2 = 60.1$
% SiO_2 in magnesium olivine $= 42.7\%$

So $5\,t\,Mg_2SiO_4$ contains $2.14\,t\,SiO_2$.

Relative formula mass of $CaAl_2Si_2O_8$ ($= CaO.Al_2O_3.2SiO_2$) $= 278.3$
% SiO_2 in calcium plagioclase $= 43.19\%$

So $5\,t\,CaAl_2Si_2O_8$ contains $2.16\,t\,SiO_2$.

$90\,t$ of remaining magma contains $(49 - 4.3)\,t\,SiO_2$
$\qquad\qquad\qquad\qquad\qquad\qquad = 44.7\,t\,SiO_2$
So the percentage SiO_2 in the remaining magma $= 49.67\%$.

(ii) The magma contains 49% SiO_2. The olivine and plagioclase both contain 43% SiO_2, so when they crystallize they remove proportionately less SiO_2 than other oxides. This causes the percentage of SiO_2 to increase in the remaining liquid.

(c) Increase in SiO_2, Na_2O and K_2O from basalt through andesite to rhyolite. Other oxides show a decrease.

(d) Andesites become richer in K_2O as the distance above the Benioff zone increases (see Fig. A3.5).

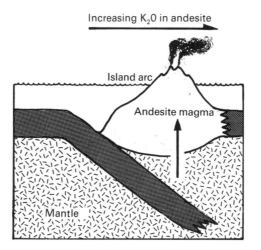

Fig. A3.5

There are two ways in which this relationship might arise.
1 Magmas are thought to come from the mantle wedge above the Benioff zone. The relationship indicates that the degree of partial melting of peridotite decreases from the Benioff zone into the mantle wedge.
2 The relationship may reflect increased degrees of magmatic differentiation experienced by magmas derived from just above the descending plate. In areas of thicker mantle, magmas have a longer time for differentiation before reaching the surface.

3.6 (a) (i) Zones I and IV are chilled margins. On coming into contact with the country rock, the magma cools quickly so only small crystals can grow.

(ii) An igneous rock formed by the settling of early-formed crystals to the base of a magma chamber.

(iii) The dense magnetite has crystallized early and sunk to the floor of the magma chamber.

(iv) From Zone I or IV. Here, crystallization has occurred before gravity settling of crystals produced compositional variation in the middle of the sill.

(b) (i) For a plagioclase crystal with a radius of 0.5×10^{-3} m:

$$\text{settling velocity} = \frac{2 \times 9.81 \times (0.25 \times 10^{-6}) \times (0.1 \times 10^3) \text{ ms}^{-1}}{9 \times (3 \times 10^2)}$$

$$= 1.82 \times 10^{-7} \text{ ms}^{-1}$$

Settling velocity in metres per year $= 1.82 \times 10^{-7} \times 31\,536 \times 10^3$
$$= 5.7 \text{ m yr}^{-1}$$

Similarly, to the nearest metre, other settling velocities are shown in Table A3.6(b)(i).

Mineral	Plagioclase		Olivine		Pyroxene		Magnetite	
Crystal radius (m × 10^{-3})	0.5	1.0	0.5	1.0	0.5	1.0	0.5	1.0
Settling velocity (m yr^{-1})	5.7	23	64	256	40	160	150	600

Table A3.6(b)(i)

(ii) See Table A3.6(b)(ii)

Mineral	Plagioclase		Olivine		Pyroxene		Magnetite	
Crystal radius (m × 10^{-3})	0.5	1.0	0.5	1.0	0.5	1.0	0.5	1.0
Settling time (years)	158	39	14	3.5	22.5	5.6	6.0	1.5

Table A3.6(b)(ii)

(iii) Growth or solution of crystals during sinking; convection in the magma; development of zones of various density or viscosity in the magma.

3.7 (a) See Fig. A3.7.

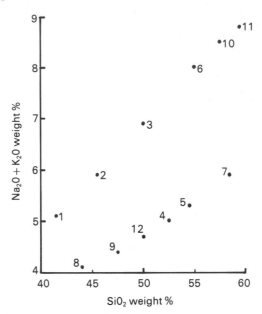

Fig. A3.7

(b) Two. The plots of $Na_2O + K_2O$ against SiO_2 show two distinct trends.
(c) They are directly proportional.
(d) C, E, F.

3.8 (a) The volume of a cone is $\frac{1}{3}\pi r^2 h$. The cone has a volume of $18.85\ km^3$.
(b) The area of an ellipse is πab, where a is the short radius and b is the long radius.
The volume of ash on the ground is $12.57\ km^3$.
Since the volume of ash lost to the atmosphere is $1.2\ km^3$, the total ash volume is $13.77\ km^3$.
(c) The volume of ash not accounted for is $5.08\ km^3$.
(d) This material sank into the magma chamber under the volcano.
(e) A caldera.

3.9 The mass of ejected material, its temperature, specific heat capacity and the height to which it was thrown. From this you could find the heat energy and the potential energy. You would also have to find the amount of energy which went into moving surrounding rocks, water and air.

3.10 (a) A few days before the eruption, the rate of tilting speeds up. Just before the eruption, there is a sudden reversal in tilting.
A few days before the eruption, rising magma inflates the inside of the volcano like a balloon. When air enters the nozzle of a balloon on its way out, the walls of the balloon deflate. Similarly, just before the eruption, the magma rises up the volcanic pipe and the magma chamber deflates. The eruption follows very soon after deflation begins.
(b) (i) Granodiorite.
(ii) The lava is extremely viscous. (Dacite is a million times more viscous than Hawaiian basalt.)

(iii) The eruption took place on May 14, 1982. (Fig. A3.10(b))
Just before the eruption, the rate of change of distance increases because the dome is swelling rapidly as magma is injected into it.

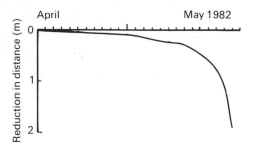

Fig. A3.10(b)

(c) (i) Rising magma pushes parts of the crater floor up and out from the vent towards the rigid crater walls.
(ii) Eruptions took place on September 6 and October 30, 1981.
Just before an eruption the rate of change of distance increases. Magma being intruded into the volcano produces rapid movement on the thrusts (Fig. A3.10(c)).

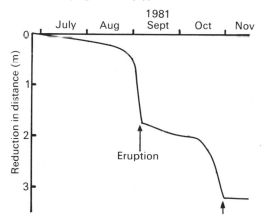

Fig. A3.10(c)

(d) (i) Seismogram 1: Produced by movements on distant faults.
Seismogram 2: Produced by movement of magma under the volcano.
Seismogram 3: Produced by tephra falls, rockfalls, snow avalanches and gas escapes.
Seismogram 4: The source of this continuous vibration is in doubt. It has been suggested that it is caused by the movement of magma through cracks. It may also be an effect like water hammer in domestic water pipes. (This is caused by sudden stoppage of flow.) Perhaps, too, gas in the volcanic pipe vibrates like an air column in a church organ pipe.
(ii) 1 Magma moving to high levels in the volcano before the eruption.
2 Magma working its way up the inside of the volcanic pipe produces numerous earthquakes so total seismic energy increases just before the eruption.

3 When the eruption begins, internal pressure has been released so earthquakes cease inside the volcano. Surface events are caused by tephra falling and rolling down the side of the volcano.

4 As the magma rises to high levels in the volcano, it may provoke harmonic tremors as it moves jerkily through cracks or it may release columns of gas which begin to vibrate.

(iii) 1 The foci probably outline the position of the magma chamber which supplied the explosive eruptions. The earthquakes indicate that the magma chamber was being refilled after the eruptions.

2 The magma source for the dome-building eruption seems to lie at a high level within the volcano.

(e) 1 As magma rises to higher levels, pressure on it is gradually reduced. This allows more and more gas to escape.

2 Much of the gas escapes from fumaroles. After an explosive eruption, the rocks of the volcano are shot through with large volumes of gas which later escapes.

3 Explosive behaviour in 1980 was accompanied by the emission of large volumes of gas. The 1983 rates probably represent the steady outgassing of the magma chamber which was refilled after the 1980 eruptions.

4 As the magma approaches the surface, pressure release causes an accelerated release of gas. The eruption brings gas-filled magma to the surface so rates of release remain high. After the eruption outgassing of the magma chamber continued as before.

3.11 In the centre of the photograph is a volcanic cone with a distinct crater. The cone shows little sign of erosion so it is apparently of recent origin. On the left is a prominent ridge which appears to be the remains of a much older volcanic cone. The ridge is probably part of the wall of a caldera. Around the base of the cone are black, lobate lava flows. Within the crater a lava flow can be seen in section. The pale material making up the cone appears to be tephra. In places, strands of pale tephra have been deposited—probably from mudflows—on the surfaces of the lava flows.

The mottled foreground appears to consist of ancient lava flows and tephra. The older lava flows cannot be clearly distinguished. This suggests that exposure to weathering causes them to lose their dark surfaces. The dark oval mound on the left of the photograph is of problematical origin. It may be the remains of a small ancient cone formed within the caldera. The photograph, however, shows no sign of a crater.

SURFACE PROCESSES

4.1 Spheroidal or onion-skin weathering. Chemical weathering affects joint blocks from all sides. Weathering penetrates most deeply at the corners so the block becomes rounded. Shells of weathered material flake off in succession. The process is aided by expansion and contraction caused by temperature change and by frost shattering.

4.2 **(a)** (i) Mass movement is gravitational downslope movement of materials.
(ii) Heaving, continuous creep and the activities of plants and animals contribute to soil creep. Heaving is caused by wetting and drying or by freezing and thawing. Expansion pushes particles out at right angles to the slope while on contraction, the particles move vertically downwards. In continuous creep the soil flows like putty. Trampling, burrowing and root growth also promote downward movements.
(iii) Solifluction: In periglacial regions, this is the slow flow of lobes of poorly sorted, saturated debris.
Earthflow: Flow of tongue-like masses of saturated regolith for relatively short distances down fairly steep slopes.
Mudflow: Flow of stream-like masses of liquid mud often for many kilometres. Characteristic of semi-arid regions.
Rockslide: In hard rock, failure on joint, bedding or fault planes causes downhill sliding.
Landslide: Sliding of regolith or bedrock on planes of weakness such as soil horizons, permafrost boundaries, bedrock surfaces or fracture planes in weak rocks such as clay.
(iv) W: solifluction; X: soil creep; Y: rockslide; Z: mudflow.
(b) (i) See Table A4.2

Method	Advantages	Disadvantages
1 Surface markers	Movements easily measured.	Affected by wash as well as creep. Gives no information about subsurface movements.
2 Vertical surface pegs	Movements easily measured.	Pegs remain straight so subsurface movements cannot be accurately measured. Difficult to calculate volumes of soil in motion.
3 Plastic tubing	Since tubes bend, subsurface movements can be measured.	Difficult to record deformation with any degree of precision.
4 Buried pegs	Subsurface movements can be accurately measured. (Probably the best method of the four.)	Pegs may be disturbed during excavation.

Table A4.2

(ii) The buried pegs should behave exactly as the regolith does. A difference in density might cause the pegs to move more or less quickly than the regolith.

4.3 (a) Solution loss is the removal of soluble weathering products. In siliceous rocks hydrolysis may lead to loss of 10–50% of the rock mass. Limestones may be completely dissolved by carbonation.

(b) Vertical transport: 48%; horizontal transport: 49%.
Solution loss depends on processes of chemical weathering which are often thought to be of only minor importance in cold climates.

(c) The rapid processes are rockfall; slush and snow avalanche; and debris avalanche, debris slide and mudflow. In combination they are responsible for 49% of the vertical transport and 46% of the horizontal transport.

(d) The average gradient is 45° so vertical movement equals horizontal movement. On shallower slopes, the horizontal component exceeds the vertical component.

(e) The slow processes are talus creep and solifluction. In combination they are responsible for 3% of the vertical transport and 5% of the horizontal transport.

(f) Polar environments. A study in Greenland found solifluction to be the dominant process with rapid mass movements being second in importance.

4.4 (a) $2.7 \times 10^4 \, m^3$ (b) 10.8%   (c) $4050 \, m^3$ (d) $10530 \, t$ (e) $13770 \, m^3$
(f) $13770 \, t$ (g) $2700 \, t$ (h) $13230 \, t$ (i) 34%   (j) 0.29
(k) Remains of land plants washed in. Remains of plants and animals in reservoir and streams.
(l) $127 \, t$ (m) $48.8 \, m^3$ (n) $46.6 \, t$ (o) $17.9 \, m^3$ (p) $0.0179 \, mm$
(q) $0.037 \, kg \, m^{-3}$
(r) The large rivers are still moving material left by glaciers.

4.5 (a) See Table A4.5.

Current velocity ($cm \, s^{-1}$)	Particle diameter (mm)				
	40	4.0	0.4	0.04	0.004
Particles picked up	150	40	16	24	60
Particles deposited	60	20	6	0.16	0.0014

Table A4.5

(b) Sand.

(c) Clay is sticky or cohesive so the particles cannot be easily picked up. Since the particles are extremely small, they can easily remain in suspension.

(d) Sand.

(e) The bend in the graph shows that the large particles do not settle as quickly as might be expected. The small particles sink relatively quickly because they do not create turbulence in the water. When a large particle sinks, it has a turbulent wake behind it which causes drag and so slows the rate of settling.

4.6 (a) By frost shattering.

(b) Scree formation has ceased or almost stopped. The screes probably formed during or soon after the last period of glaciation when daily temperatures ranges were much more extreme than at present.

(c) The terraces are probably moraines redistributed by meltwaters when glaciation ceased. When the ice melted, the valley would have been occupied by a large braided stream.

(d) The rocks above the scree slopes show well-developed cleavage. The intersection of bedding and cleavage has cut the rock into relatively thin slices.

(e) In essence, sphericity is the cube root of:

$$\frac{a}{a} \times \frac{b}{a} \times \frac{c}{a}.$$

The degree of sphericity decreases as the axes become more dissimilar. Roundness is a measure of the smoothness of surface curvature of a pebble. An egg is not spherical but it is well-rounded.

(f) A marble and a cube have equal sphericities (sphericity $= 1$). A well-known matchbox has a sphericity of 0.47; it is much less spherical than a cube.

(g) A marble is well-rounded. A cube and a matchbox are very angular.

(h) Mostly by rolling and sliding. Pebbles may saltate (bounce) when currents are strong.

(i) Pebbles tend to become more rounded and more spherical

(j) Rocks which break into equidimensional fragments will readily become nearly spherical and well-rounded. Rocks which are strongly foliated or lineated may easily become well-rounded but they are unlikely to become nearly spherical.

(k) In the river bed, pebbles have a restricted size range because they are exposed to relatively constant currents. This means that sorting has occurred so pebbles of a similar size have ended up in the same place. The river terrace deposit is less well sorted because it has been deposited by more variable currents.

(l) The largest pebbles are from the scree (mean axis length 2.93 cm). They have the lowest degrees of sphericity (0.52) and roundness (0.17). The pebbles from the large and small tributaries are of similar size (mean axis length 2.70 and 2.74 cm respectively). Their sphericities are also similar (means 0.65 and 0.64). The pebbles from the large tributary are much more rounded (roundness index 0.53) than the pebbles from the small tributary (index 0.31). The pebbles from the main river and from the river terrace are similar in size (mean axis length 2.38 and 2.50 respectively). Their sphericities are also similar (0.66 and 0.67). The river terrace pebbles show higher degrees of roundness (0.64) than the pebbles from the main river (0.55).

Differences such as those described here may or may not be indicative of real differences between the processes acting on the pebbles. If there are no real differences between the processes or the degrees to which the processes have been acting, then the differences in pebble shape would be due to chance.

(m) Boulders, cobbles and other particle sizes have not been measured.

4.7 (a)　　(i) A: Carbonation-solution of limestone along joints forms open grikes separating blocks called clints.
B: The rock is broken by frost shattering and by solution along cracks. At the cliff base, the broken rock accumulates as scree.
C: The turf rolls have been produced by solifluction. Solifluction leaves an ill-sorted, moraine-like deposit called head.

      (ii) On falling, large particles travel further down the scree slope because they have more momentum than small particles. (Momentum increases as the cube of the particle diameter, while friction increases as the square of the diameter.)

      (iii) Gorges are formed by river erosion and by the collapse of cavern roofs.

  (b) (i) At equal water temperatures (e.g. in April and December) calcium carbonate solubility differs.

      (ii) Levels of dissolved carbon dioxide in the water from the soil.

      (iii) The carbon dioxide is produced by the respiration of soil organisms. The rate of respiratory activity increases with the warmth of summer.

      (iv) Rain water dissolves acidic oxides (e.g. CO_2, SO_2, NO_2) from the atmosphere and soil. The resulting acids react with and dissolve the limestone. Carbonation is one such reaction:

$$H_2CO_3 + CaCO_3 \longrightarrow Ca^{2+} + 2HCO_3^{-}$$

4.8 **(a)** Raised shorelines. Difficult to say because of complex interplay of late- and post-glacial eustatic and isostatic changes. Shorelines probably form when the land and sea are rising at the same rate so, for a time, they retain a fixed position relative to each other. Periods of rapid isostatic movement punctuated by times when the land and sea were rising at equal rates would form a series of shorelines.

  (b) (i) Because the main centre of ice accumulation lay to the west, there was more isostatic recovery in the west than in the east.

      (ii) The shorelines followed the retreat of the ice to the west.

      (iii) The amount of isostatic uplift has decreased between the formation of the oldest and youngest shorelines.

      (iv) The order would be reversed.
In the west, the land rose faster than the sea. In the east, the sea rose faster than the land.

      (v) 227 m.

4.9 **(a)** At position A, discs and blades predominate because they are the least streamlined shapes. During storms they can easily be carried to high levels on the beach. The frequency difference between the rods and spheres does not appear to be significant.

At C, spheres and rods are common because they are the most streamlined shapes so they are not easily picked up and carried to high levels. Such rounded shapes also have a tendency to roll to lower levels on the beach.

At B, an intermediate condition exists. Discs and blades are dominant and rods appear to be significantly more common than spheres. Spheres are more streamlined than rods so would tend to occupy lower levels on the beach.

  (b) Pebble shapes are partly determined by structures such as foliation, lineation, bedding, cleavage and jointing in the rock from which the pebbles are derived. Slate has a very well-defined cleavage so it is likely that most of the slate pebbles will be discs. Gneiss may have a well-defined foliation and lineation so the gneiss pebbles will probably be discs and blades with a few rods. The sandstone may or may not be well-bedded so it may produce pebbles of various shapes. It is likely that most sandstone pebbles will be discs with small numbers of spheres, rods and blades.

Dolerite has no foliation or lineation but it may be finely jointed. It is likely that most of the spheres will be dolerite. However, since dolerite pebbles are very numerous, other shapes will be well represented.

Overall, the slate and gneiss pebbles will probably be found at relatively high levels on the beach. Dolerite, because of its high density and tendency to form rounded shapes, may occur mostly at relatively low levels. It is difficult to predict the distribution of the sandstone pebbles. Most sandstone pebbles will probably be found at relatively high levels.

4.10 (a) (i) A kettlehole;   B drumlin;   C kame terrace;   D esker; E terminal moraine;   F recessional moraine;   G outwash plain; H kame;   I crevasse filling;   J ground moraine;   K moraine dammed lake.

A: Melting of an ice block buried in drift leaves a steep-sided depression.

B: Boulder clay is moulded into streamlined shapes by the moving ice sheet. The blunt end of the drumlin points upstream.

C: Deposited by a stream running between ice and valley wall.

D: Deposited by a stream running under the ice.

(ii) Esker sediments show bedding, current-produced structures, water-worn pebbles and signs of sorting. Morainic sediment is unsorted and it has little internal structure. The pebbles and boulders are usually angular.

(iii) Varves are annual layers deposited in proglacial lakes. Silt deposited in summer is succeeded by clay deposited in winter when the lake surface is frozen. Counting and correlating varves has produced a detailed chronology of the last 17 000 years.

(b) (i) Contributory processes may have included: 1 Differential loading by pebbles;   2 On freezing, the wet sand would expand more than the pebble layer—this may have set up stresses which deformed the sand;   3 The newly-deposited sand while still in a semi liquid state may have been acted on by currents; and   4 The sediment may have been pushed by ice or it may have lost the support of a melted ice wall.

(ii) Climbing ripples result from high rates of deposition by decelerating currents. In this case the current flowed from right to left.

(iii) The boulder, encased in a block of ice, has been washed into a lake or stream eddy. Slow sinking as the ice melted gently deposited the boulder.

(c) (i) Patterned ground is formed under arctic or sub-arctic conditions.

(ii) Freezing causes the ground surface to contract and form mud polygons. Freezing and thawing causes stones to move upwards and, on reaching the surface, they move down the sloping polygon surfaces to collect at the polygon margins. On slopes, the circles are drawn out to form stripes.

(d) (i) Different levels of precipitation. Snowfall in the west exceeded that in the east. The larger volumes of ice formed in the west would survive at lower altitudes in summer or during times when the climate was relatively warm.

Differences in temperature. Corries can form at lower altitudes in colder climates. This is unlikely to have been a factor here since temperatures probably decreased from Skye to the Cairngorms. Corries also form at lower altitudes on colder, north-facing mountain slopes. There is no evidence here that north-facing and south-facing corries are being compared.

(ii) When snow in a hollow reaches a thickness of about 30 m, its base is transformed into ice. Excess accumulation of ice against the back of the hollow coupled with ablation at the mouth causes the ice to flow with a rotational motion. The hollow is enlarged by abrasion and quarrying (plucking) to become a cirque. The steepness of the headwall is maintained by frost shattering. See Fig. A4.10.

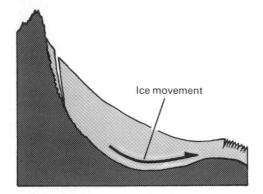

Fig. A4.10

4.11 (a) (i) A canyon; B butte; C barchan dunes; D playa lake; E alluvial fan.

(ii) Barchan dunes tend to form where there is a limited supply of sand. The dune moves in the same direction as the wind because sand is blown from the shallow upwind slope to accumulate on the steep downwind slope. Barchan dunes form under the effects of wind blowing consistently from one direction. They have crescentic forms with their concave sides downwind. Longitudinal dunes form under the action of winds which blow from two dominant directions, so the dunes are parallel to the average wind direction. See Fig. A4.11.

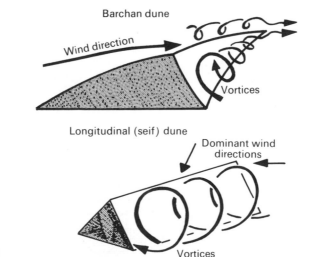

Fig. A4.11

(iii) Mass movements which lower slopes are much slower in arid regions.
(iv) Clay and salts.

(b) (i) The growth, thermal expansion and hydration of salt crystals within the rock produce stresses which break the rock.

    (ii) Factors might include:

      1 Solubility. When a rock is dried out, a very soluble salt will leave behind a large number of crystals while a sparingly soluble salt will leave few crystals.

      2 The degree of hydration of the crystals. Because of its greater volume, a hydrous crystal can exert a greater pressure than an anhydrous crystal. Note that there is no simple relationship between the number of water molecules in the crystal and the ability of the salt to break the rock, e.g. $Na_2CO_3.10H_2O$ has little effect on the sandstone while $Na_2SO_4.10H_2O$ produces rapid disintegration.

      3 The form of the crystal. Some crystals may have feathery forms which exert little internal pressure on the rock. Others may have compact forms which exert considerable pressure during growth.

    (iii) Crystallization of salt within the cube causes an increase in weight.

(c) (i) Sodium sulphate is the salt which produces most rapid rock disintegration.

    (ii) Chalk absorbs a great deal of salt solution so numerous crystals grow within it. It has little internal strength so it is easily broken up. Granite is impermeable so it does not absorb salt solution. Being crystalline, it is also very strong.

    (iii) The reason cannot be certainly stated. Possibly because repeated stressing over previous days eventually fractured the iron oxide cement and caused sudden failure of the outer layers of the cube.

SEDIMENTARY ROCKS

5.1 1 Pyroclastic sediments; 2 detrital sediments; 3 chemical precipitates; 4 $CaCO_3$ and SiO_2 from shells, corals, sponges, radiolarians, etc.; 5 carbon and hydrocarbons.

5.2 **(a)** Sandstone consists largely of quartz (SiO_2). Besides quartz, shale contains minerals such as feldspar and clay, which are partly SiO_2, and minerals such as calcite which have no SiO_2.

(b) Shale contains many minerals (e.g. feldspar, clay, mica, chlorite) which consist partly of Al_2O_3. Sandstone contains few of these minerals.

(c) The silicate minerals of igneous rocks have most of their iron as Fe^{2+}, so FeO predominates over Fe_2O_3. On weathering, Fe^{2+} is readily oxidized to insoluble Fe^{3+} which accumulates in sedimentary rocks. Much Fe^{3+} occurs in shale as hydrated oxides such as limonite. The Fe^{2+} as FeO in shale may occur in minerals such as biotite, chlorite and siderite. Pyrite forms under reducing conditions.

(d) On the weathering of silicate minerals, most of the Mg^{2+} and Ca^{2+} go into solution. Some Mg remains to be deposited in shale as biotite and chlorite. Some Ca may be found in detrital plagioclase but most Ca will be found in authigenic calcite and dolomite.

(c) On weathering, Na^+ is removed in solution. The Na in the sedimentary rocks is probably in detrital plagioclase.

(f) K occurs in shale as a clay mineral called illite. There is no equivalent Na clay mineral.

(g) Hydrated minerals (e.g. clay, chlorite, limonite) are weathering products commonly found in shale. OH^- is also present in micas.

(h) The sandstone probably contains a small amount of intergranular calcite. Calcite, dolomite, siderite and carbonaceous material may be present in the shale. Carbonates do not commonly crystallize from magmas.

5.3 (a) A deposition; B diagenesis; C sedimentary rocks; D heat and pressure; E metamorphic rocks; F melting; G solidification; H igneous rocks; I rocks moved to surface; J denudation; K transport.

(b) Diagenesis: Chemical and physical processes affecting a sediment at or near the Earth's surface.

Lithification: Processes, including compaction and cementation, by which a soft sediment is converted to a hard rock. Part of diagenesis.

Authigenic minerals: Minerals grown in their place of occurrence. Minerals formed with or after the rock of which they form a part. Usually applied to minerals formed in sediment during or after deposition; e.g. growth of authigenic quartz on quartz sand grains; growth of calcite on shell fragments.

(c) Connate water is water trapped in sedimentary rocks during deposition. The rock is partly dewatered by increased pressure during burial. Juvenile water is magmatic water which has come to the Earth's surface for the first time.

5.4 (a) See Fig. A5.4(a).

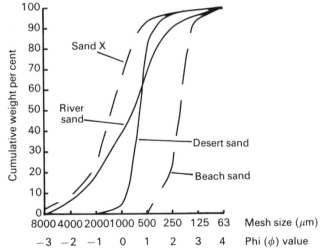

Fig. A5.4(a)

(b) (i) In a well-sorted sediment most of the grains are of a similar size. In a poorly-sorted sediment the grains are more evenly distributed over a range of sizes.

(ii) Approximate sorting coefficients: River sand 1.30; desert sand 0.43; beach sand 0.45; Sand X 1.18.

(c) (i) In the upper course of a river, the fine-grained material is washed away to leave a deposit which has an excess of coarse-grained particles. Such a deposit may have a negative skewness. In the lower course of the river, deposition of the fine-grained material produces a deposit with an excess of such particles. This sediment may have a positive skewness.

(ii) Approximate skewness values: River sand −0.20; desert sand 0.18; beach sand 0.10; Sand X 0.13.

(d) See Fig. A5.4(d).

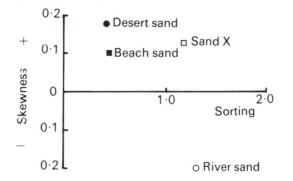

Fig. A5.4(d)

(e) The desert sand has well-rounded grains frosted by collisions in air. The winnowing action of wind has produced a well-sorted sediment. The positive skewness may be the result of the deposition of smaller grains among the larger grains during times of reduced wind strength.

The beach sand consists of grains rounded and polished by being moved up and down the beach by waves. It has been well-sorted by constant wave action. The slight positive skewness is indicative of a low-energy environment where an excess of fine-grained material has been deposited.

The river sand has grains showing some effects of rounding and polishing by water transport. It is poorly-sorted because the inconstancy of river currents means that variously-sized particles are deposited in the same place. The negative skewness suggests that fine-grained material has been washed downstream leaving a predominance of large particles.

The very angular grains of Sand X suggest a glacial environment of deposition while the polish on some of the subrounded grains indicates water transport. The poor sorting suggests that the sand has been deposited by a current suddenly slowed. Sand X has probably been deposited by meltwater from a glacier perhaps in an outwash fan or in the delta of a proglacial lake. The positive skewness indicates that fine-grained material from the glacier was deposited with the coarser material instead of being washed away as it could be in a river channel.

To allow you to define more certainly the environment of deposition of Sand X, much more information is required: e.g. field evidence; analysis of many more samples; and knowledge of the area from which the sediment was derived, because a sediment may inherit the characteristics of pre-existing rocks.

5.5 These are glaciofluvial deposits. The regularly-bedded lower layer has a pebbly base and a sandy top. The top of the lower bed has been eroded away perhaps when currents changed direction and speeded up. The base of the upper layer shows cross-bedding, indicating that the water flowed from right to left. Large, water-worn pebbles above the cross-bedded unit show that currents were strong. Signs of cross-bedding are just visible in the pebbly layer. As the currents slowed, a sandy layer was deposited. Pebbles reappear in the coarse-grained sand at the top of the photograph.

5.6 The features are groove casts on the underside of the greywacke bed. Turbidity currents drag pebbles and boulders over the surface of sea-floor mud. The grooves which are cut are filled by turbidite material so the base of the bed deposited by the current exhibits casts of the original grooves.

5.7 (a) Initially, limestones were deposited over the whole area of a small subsiding basin. With continued subsidence, the basin deepened. Shales were deposited in the deep offshore areas while limestones were deposited in shallow water near the edge of the basin. With further subsidence, a river from the north left deltaic deposits while in the south, offshore siltstones were being deposited. Finally, the delta continued its southerly advance and the siltstones were covered by deltaic sandstones.

(b) Coral reefs grow best in well-lit water down to depths of about 20 m. Only around the edge of the basin would the water have been shallow enough to allow coral growth.

5.8 (a) Area A: alluvial fan gravels and sands.
Area B: wind-blown sand and playa clays and salt deposits.
Area C: wind-blown sand (perhaps finer grained than that of area B) and possibly loess. May be subsurface gypsum.
Open sea: Sand from the land and wind-blown dust which would settle as clay. Coral reefs may well be present.

(b) Partly enclosed basin: Wind-blown sand would be found in the north-west part of the basin while clay formed from dust would occur in other areas. Evaporites will probably be forming since rates of evaporation will be high.

(c) When about half of a given volume of sea water has evaporated, $CaCO_3$ begins to precipitate. When the volume has been reduced to about 20% of the original, $CaSO_4$ begins to precipitate. When 10% remains, NaCl begins to precipitate while very soluble K and Mg salts precipitate from the last 4% of the sea water. The sequences in most evaporite deposits show that evaporation of sea water has been partial rather than complete. Most evaporite deposits have been formed by evaporation from subsiding lagoonal basins which received frequent supplies of salt water. Thick evaporite deposits cannot have been formed by the total evaporation of seas. For example, if the North Sea evaporated, the salt in most places would be only about 2 m thick.

5.9 (a) One sequence is the opposite of the other. The sequence in Fig. 5.9A are deposits formed as the sea moved over the land from east to west. The sequence in Fig. 5.9B was formed as the sea retreated from the land from west to east.

(b) Different parts of a diachronous ('across time') bed have different ages. On any horizontal surface in Fig. 5.9, the ages of the various sediments are approximately the same. Any one unit is diachronous because it gets younger from depth towards the surface.
Facies ('general appearance') describes sedimentary environments and the sediments deposited therein. Since environments persist through time, rocks of different ages deposited in similar environments resemble each other. The characteristics of the sediment reflect the environment of deposition. For example, the peat, silt and mud would be described as being of swamp and lagoonal facies.

(c) Cross bedding would be prominent in the sand deposited in the offshore bars. Such bars are built by wave, current and wind action so the sand is often laid in sloping layers. Cross bedding on a smaller scale will also be found in the flood plain deposits.

Mud cracks would be found in the flood plain and in the swamp and lagoonal deposits. From time to time sediment in these environments will be desiccated on exposure to the air.

EARTH PHYSICS

6.1 (a) Density: As rock density increases, the velocities of both waves decrease.
Compressibility: As rock becomes more compressible the velocity of P-waves decreases. Conversely, as the rock becomes more incompressible, the velocity increases.
Rigidity: As rock becomes more rigid the velocity of S-waves increases. S-waves do not pass through liquids because liquids have zero rigidity.

(b) 1 The gradual increase in density with depth is more than offset by increases in incompressibility and rigidity.

2 The waves enter peridotite which, although denser than crustal rocks, is much more incompressible and rigid.

3 In this zone (the low speed or low velocity layer) the peridotite is thought to be about 5% molten. This makes the rock more compressible and less rigid. The slight fall in density is more than offset by the changes in compressibility and rigidity.

4 Increased pressure probably produces phase changes in which minerals change to high-density, low-volume forms. Increased incompressibility and rigidity more than compensate for the increased density.

5 S-waves are not transmitted because the outer core has liquid properties and zero rigidity. P-wave velocity falls because there is a large increase in density from peridotite to liquid metal and a large decrease in incompressibility.

6 The P-waves enter the denser but much more incompressible solid inner core.

6.2 (a) (i) A:670 km; B:1330 km; C:1980 km; D:2670 km.
(ii) 3.13 km s^{-1}.
(iii) See Fig. A6.2(a)(iii).

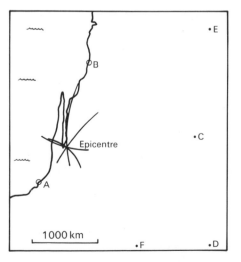

Fig. A6.2(a)(iii)

(iv) Recording station E is 2850 km from the epicentre. The P-waves would take 491 seconds (s) to arrive. Station F is 1850 km from the epicentre, so P-waves would take 319 s to arrive.

(v) The S–P time intervals are: A: 99 s   B: 196 s   C: 290 s   D: 392 s. See Fig. A6.2 (a) (v).

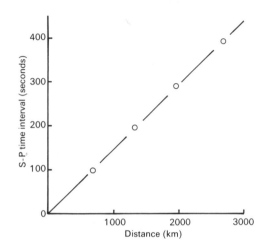

Fig. A6.2(a) (v)

The S–P time lag is 145 s per 1000 km.

(b) The epicentral distance, S–E, is less than the distance between the focus and the seismometer. The circle of radius S–F does not pass through the epicentre. A similar effect occurs with the other seismometers. See Fig. A6.2 (b).

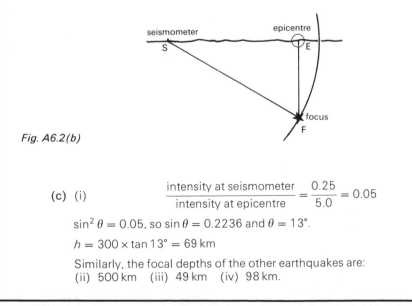

Fig. A6.2(b)

(c) (i) $\dfrac{\text{intensity at seismometer}}{\text{intensity at epicentre}} = \dfrac{0.25}{5.0} = 0.05$

$\sin^2 \theta = 0.05$, so $\sin \theta = 0.2236$ and $\theta = 13°$.

$h = 300 \times \tan 13° = 69$ km

Similarly, the focal depths of the other earthquakes are:
(ii) 500 km   (iii) 49 km   (iv) 98 km.

(d) (i) In line 1 of the table:

$$\frac{\sin \theta_I}{\sin \theta_R} = \frac{V_I}{V_R},$$

so $$\sin \theta_I = \frac{V_I}{V_R} \sin \theta_R$$

$$= \tfrac{7}{8}$$

and $$\theta = 61°$$

Similarly, in line 2 the angle of refraction is 38°; in line 3 the velocity of the refracted wave is 8.09 km s^{-1}; in line 4 the velocity of the incident wave is 8.09 km s^{-1} (from line 3) and the angle of refraction is 41°.

(ii) Slowly; quickly; quickly; slowly.

(iii) When a wave from a low velocity layer enters a high velocity layer, the wave is refracted away from the normal. A wave incident at the critical angle does not enter the high velocity layer because the angle of refraction is 90°, so the wave travels parallel to the boundary between the layers. The angle of incidence in line 1 of Table 6.2B is a critical angle.

(iv) The angle of reflection is the angle between the normal and the reflected wave. The angle of reflection equals the angle of incidence.

(v) It will be reflected.

6.3

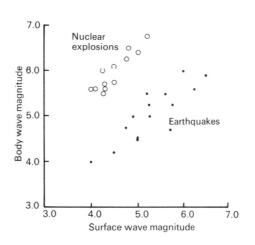

Fig. A6.3

(a) Explosions: no S-waves; originate at a point; radial pattern of wave propagation; first P-waves arriving at seismometers recorded as compressions; depth to 2 km; not located under sea floor; usually from aseismic zones.

Earthquakes: P- and S-waves; source area may be large; petal-like pattern of wave propagation; first P-waves arriving at seismometers may be compressions or decompressions; depth to 700 km; may originate under sea floor; usually from seismic zones; longer duration than explosions; usually greater magnitude than explosions.

(b) Body waves (P- and S-waves) penetrate deep within the Earth. Surface waves (L-waves), like waves in the sea, are restricted to surface layers.

(c) Explosions have high body wave magnitudes but low surface wave magnitudes (approximate ratio 1.3:1). Earthquakes tend to have lower body wave magnitudes than explosions but higher surface wave magnitudes. (In earthquakes the ratio of the magnitudes is about 1:1.)

(d) Soft, porous materials absorb the energy of the explosion much more efficiently than rigid rocks. A 20 kilotonne explosion in granite would produce a seismometer signal like that produced by a magnitude 5 earthquake. The same explosion in alluvium would produce a signal like a magnitude 4 earthquake.

6.4 (a) From the asteroid belt between Mars and Jupiter.

(b) Irons are easily recognized. Stones look like terrestrial rocks so they are generally overlooked.

(c) Some achondrites resemble basalt.

(d) The very slow rate of cooling. (The Earth's core is thought to be cooling at a rate of about 0.25 °C per million years.)

(e) It is most unlikely that the meteorites came from separate minor planets of iron, stony iron or stony composition. The meteorites probably came from different layers within the planetesimals.

(f) Irons would come from the metallic core. Most achondrites would come from the peridotite mantle. Stony irons would have been derived from the core-mantle boundary. Chondrites would come from unmelted surface layers.

(g) There is a one-to-one relationship showing that the solar abundance is the same as that in the carbonaceous chondrites. This suggests that carbonaceous chondrites were formed from the nebula at the same time as the Sun and that these chondrites have not been modified by later changes. Carbonaceous chondrites apparently provide samples of original nebular material.

(h) Chondrites provide radiometric ages of 4600 million years (Ma). This is apparently the age of the solar system and of the Earth. The Earth probably formed from chondritic material. Analysis of chondrites may give an indication of the overall chemical composition of the Earth. The Earth's core probably consists of material like that of iron meteorites. The mantle apparently consists of material like that of common achondrites. Carbonaceous chondrites contain complex carbon compounds (e.g. sugars, amino acids). This shows that organic compounds could have been produced on Earth before the appearance of living things. Such compounds may have formed the building blocks of the first organisms.

6.5 (a) 1 According to Airy's hypothesis, mountains float in the mantle like wooden blocks. The excess mass of the mountain range is compensated by a mass deficiency in the mountain root.

2 According to Pratt's hypothesis, mountains have a lower density than the rest of the crust so they float at higher levels than the rest of the crust. Seismic studies have shown that high mountains usually lie on thick crust, so Airy's hypothesis seems to apply to crust. For the lithosphere, Pratt's hypothesis seems to apply. At a depth of about 100 km the lithosphere seems to float on the asthenosphere.

(b) The ridges are in isostatic equilibrium so they are not higher because of the presence of excess material. Instead, the rocks have a relatively low density because expansion due to heating has increased their volumes.

(c) The mass of the magma column is balanced by the mass of the rock column. Where h is the height of the magma column:

$$h \times 2.7 = (43 \times 3.3) + (7 \times 2.9)$$

and $h = 60.07$ km.
The volcanic cone could reach 10.07 km above the sea floor and remain in isostatic equilibrium.

(d) 48.5 km.

(e) A positive anomaly. The volcano would sink till isostatic equilibrium had been attained.

(f) A seamount is a submarine mountain rising more than 915 m above the sea floor. A guyot (or tablemount) is a flat-topped seamount. An atoll is a ring-like coral reef.
Seamounts are volcanic cones. A guyot is a submerged volcanic island planed flat by wave erosion as it subsided. Atolls form as corals grow to keep pace with the downward movement of sinking volcanic islands.

(g) As the oceanic crust spreads away from ridges, it cools and contracts so the depth of the ocean increases. Islands formed near the ridge are carried to lower levels by the contracting crust.

6.6 (a) See Fig. A6.6.

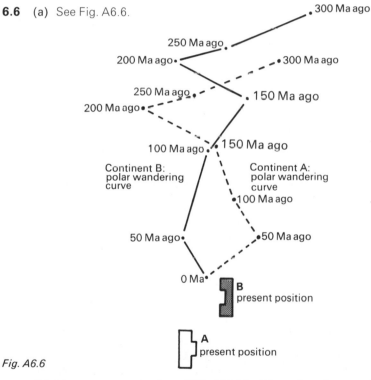

Fig. A6.6

(b) The curves overlap from 300–150 Ma. Thereafter, they are divergent.

(c) The continents wander, not the poles.

(d) Rocks of the same age on different continents give different palaeopole positions, so giving different continents different polar wandering curves. Since there can have been only one North Pole at any one time, movements of the continents must have produced the multiplicity of palaeopole positions.

6.7 (a) See Fig. A6.7.

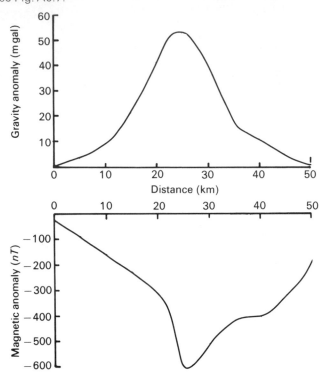

Fig. A6.7

(b) A body of high density in the crust.
(c) Continental crust has a density of about $2.7 \, g \, cm^{-3}$. The density of the body is about $3.0 \, g \, cm^{-3}$.
(d) The body is relatively rich in iron minerals. It was magnetized when the Earth's field was reversed.
(e) A gabbroic intrusion.
(f) The central area of the intrusion has a high level of remanent magnetism. It is probably richer in iron minerals.

6.8 (a) (i) By convection in the outer core and mantle and by conduction through the lithosphere. Igneous processes also carry heat to higher levels.

(ii) Heat is produced by the decay of radioisotopes such as those of U, Th and K. Such isotopes have their highest concentrations in rocks of the continental crust and their lowest concentrations in mantle rocks.

(iii) Heat from kinetic energy of infalling particles during accretion of Earth. Heat from short-lived radioisotopes which decayed soon after the Earth formed. When the core was being formed, the sinking of dense material released heat. Solidification of the liquid outer core will also release heat. Compression of material at depth generates heat (adiabatic heating).

(b) (i) 52.8% of heat escaping from continents comes from mantle.
91.9% of heat escaping from ocean floor comes from mantle.

(ii) The mantle under the oceans may be hotter because convection currents rise under the ridges and move off sideways or because this part of the mantle contains more radioisotopes. If the mantle under the oceans and continents is at the same temperature, the mantle under the crust may be better insulated by a thicker lithospheric layer.

(iii) Oceanic crust is initially very hot and the rate of cooling of a hot body is proportional to the temperature difference between it and its surroundings. Also, oceanic crust has few radioisotopes to help maintain its high temperature. Continental crust has relatively high concentrations of U, Th and K which have long half-lives, so heat flow falls slowly.

METAMORPHIC ROCKS

7.1 (a) migmatite (b) slate (c) hornfels (d) phyllite (e) serpentinite (f) granulite.

7.2 (a) Granoblastic texture. Equigranular and equidimensional grains form an interlocking mosaic with the grains meeting at angles of about 120°. In three dimensions, the grains would resemble a mass of soap bubbles. Often forms in hornfels because quartz grains readily form an interlocking mass. Texture resembles that of an annealed metal so it probably forms in an environment of low directed stress.

(b) Porphyroblastic texture formed by the growth of large crystals in a finer-grained matrix (e.g. garnet in schist; augen of feldspar in gneiss; andalusite in hornfels). Minerals slow to nucleate tend to form a few large crystals while minerals which nucleate easily form large numbers of small crystals.

(c) Porphyroclastic texture formed when cataclasis leaves relatively large unbroken fragments in a finer-grained matrix. Common in mylonites.

(d) Sedimentary layering has been folded into the form of similar folds. Deformation and crystallization of quartz, feldspar and mica have formed an axial plane foliation.

(e) Garnet grew over an early, folded foliation. Later metamorphism under directed stress produced a new foliation. However, the garnet did not recrystallize so the early foliation was preserved within it.

(f) This is partly metamorphosed gabbro. The olivine and plagioclase are stable at the temperature of formation of the gabbro but they are unstable during metamorphism at lower temperature. During metamorphism the olivine reacts with the plagioclase to form a corona or reaction rim of hornblende. During the reaction Mg and Fe diffuse out of the olivine and Ca diffuses in.

7.3 (a) (i) S1: Amphibolite or hornblende-plagioclase schist. The well-defined foliation shows it to be metamorphic.
S2: Olivine dolerite. Grain size and interlocking igneous texture in which minerals show no preferred orientation.
R: Quartz–feldspar gneiss. Pale colour and banded nature in photograph. Grain size and preferred orientation of minerals.

(ii) After the dolerite had been intruded and cooled, it was subjected to metamorphism such that the margin of the intrusion recrystallized as amphibolite while the interior of the intrusion remained unchanged.

(b) Gneiss was formed probably from metamorphosed sedimentary rocks. One or more small dolerite dykes or sills were intruded into the gneiss. Compression at a high angle to the gneissose banding folded the intrusions. Stretching parallel to the banding pulled the intrusions apart. The dyke margins were probably metamorphosed during folding when temperatures were relatively high.

7.4 **(a)** (i) 1 shale; 2 spotted shale; 3 chiastolite shale; 4 calcium silicate hornfels; 5 hornblende plagioclase hornfels; 6 granite; 7 tourmaline granite.

Spotted shale: Under the effects of low-grade contact metamorphism, ore minerals and organic matter coalesce during recrystallization as spots of magnetite or graphite.

Chiastolite shale: With increasing intensity of metamorphism, andalusite crystals grow in the shale. Chiastolite is a form of andalusite with an internal cross of inclusions. The inclusions disappear as the grade of metamorphism increases.

Calcium silicate hornfels: Produced by the metamorphism of an impure limestone. Calcite reacts with impurities such as Si, Al, Mg and Fe to produce a rock consisting of Ca silicates such as pyroxene, amphibole, plagioclase, garnet and epidote.

Hornblende plagioclase hornfels: Produced by the metamorphism of a basic igneous rock such as dolerite. The augite of the dolerite recrystallizes as hornblende.

(ii) Stream-tin: Also called pebble-tin. Weathering and erosion of mineral veins produces grains and pebbles of cassiterite which collect in river gravels.

Tor: A prominent, isolated, hill-top exposure of jointed granite.

Sheet joints: Joints in plutonic igneous rocks formed parallel to the ground surface by pressure release as the rock expands and splits with the removal of overlying rocks by denudation.

(b) Following deposition of limestones, shales were deposited. The sedimentary rocks were intruded by a dolerite sill. These rocks were metamorphosed by the intrusion of granite. At a late stage of intrusion, parts of the granite were apparently affected by boron-rich pneumatolytic fluids which replaced mica with tourmaline to form tourmaline granite. Apparently, too, hydrothermal cassiterite deposits are genetically related to the granite. Later, deep weathering of the granite along joint planes led to the formation of tors. Eventually, placer deposits of cassiterite were left by streams.

7.5 **(a)** Metamorphic grade: A description of the intensity of metamorphism. Metamorphism at high temperature would be described as high grade while metamorphism at low temperature would be described as low grade.

Index mineral: A distinctive mineral which appears in a rock when a certain grade of metamorphism is reached.

Metamorphic zones: Areas of differing metamorphic grade recognized by the successive appearance of diagnostic index minerals. Zones are most easily established in metamorphosed argillaceous rocks.

Isograd: A line on a map separating different metamorphic zones. The line joins points of equal metamorphic grade.

(b) (i)

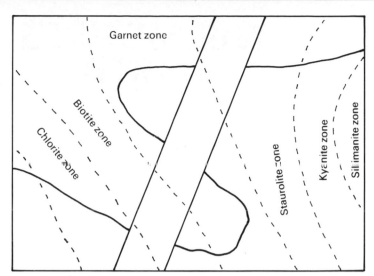

Fig. A7.5

(ii) The grade of metamorphism increases from the chlorite zone towards the sillimanite zone; i.e. from W to E.

(c) P: chlorite, quartz, plagioclase, mica.
Q: biotite, chlorite, plagioclase.
R: garnet, calcite, epidote, hornblende.
S: garnet, epidote, plagioclase, hornblende.
T: garnet, hornblende, plagioclase.
U: sillimanite, garnet, mica, quartz, plagioclase.
Rocks of differing chemical composition subjected to the same grade of metamorphism produce different mineral assemblages.

7.6 **(a)** 4 km; kyanite   (ii) 12 km; kyanite

(b) (i) Barrow-type metamorphism has taken place at high pressure. Buchan-type metamorphism has taken place under low pressure conditions.

(ii) $0.05\,°C\,m^{-1}$; 1.67 times the present normal geothermal gradient.

(iii) $0.2\,°C\,m^{-1}$; 6.67 times the present normal gradient.

(iv) High directed pressure and friction produced during mountain building would generate heat. May have been relatively high levels of radioisotopes.

(v) It has been suggested that extra heat was brought into the orogenic belt by mantle convection. Buchan area may have been intruded by a very large high-temperature intrusion.

7.7 **(a)** A metamorphic facies is a group of mineral assemblages formed in different rocks subjected to generally similar conditions of pressure and temperature.

(b) (i) Granulite facies: 3–11 kilobars; 725–875 °C.
Blueschist facies: 3–12 kilobars; 90–250 °C.

(ii) A hornfels; B amphibolite; C eclogite.

(iii) B, E, F.

(c) Under conditions of changing fluids and fluid pressures, mineral assemblages belonging to a particular facies may be produced well outside the defined pressure and temperature conditions. The pressure and temperature conditions under which a metamorphic assemblage formed may be much broader than those suggested by the facies diagram.

THE MOVING EARTH

8.1 **(a)** Since the volume of a pebble does not change during deformation, we can say:

$$a_2 b_2 c_2 = a_1 b_1 c_1$$
$$= 1.43$$

where $a_1 b_1 c_1$ are the pebble axes before strain and $a_2 b_2 c_2$ are the pebble axes after strain.

For the quartzite pebbles in sample B:

$$a_2 = 33c_2 \quad \text{and} \quad b_2 = 25c_2$$

Therefore,

$$33c_2 \times 25c_2 \times c_2 = 1.43 \quad \text{and} \quad c_2^3 = \frac{1.43}{33 \times 25}$$

from which $c_2 = 0.12$; $a_2 = 3.96$; $b_2 = 3.0$

Strain parallel to $a_2 = \dfrac{3.96}{1.3} - 1 = 2.05$

Strain parallel to $b_2 = \dfrac{3.0}{1.1} - 1 = 1.73$

Strain parallel to $c_2 = \dfrac{0.12}{1.0} - 1 = -0.88$

Similarly, the strains suffered by the other pebbles are shown in Table A8.1.

| | Strain parallel to axis | | |
	a_2	b_2	c_2
Sample B quartz pebbles	1.52	1.22	−0.82
Sample C quartzite pebbles	2.36	2.47	−0.91
quartz pebbles	1.45	1.44	−0.83

Table A8.1

(b) The short axes of the deformed pebbles are perpendicular to the bedding. In the lower, non-inverted limb of the fold the strain at right angles to the bedding is −0.88. The ratio of present to original thickness is 0.12, so a bed originally 10 m thick would now be 1.2 m thick. In the upper, inverted limb, the strain at right angles to the bedding is −0.91. The ratio of present to original thickness is 0.09, so a bed originally 10 m thick would now be 0.9 m thick.

(c) The fact that the quartzite and quartz pebbles have different strains shows that the amount of strain depends on the properties of the material being strained. Since the pebbles are embedded in what is probably a weaker matrix, the bedding may have suffered more strain than the pebbles.

8.2 The rock shows columnar jointing. The basalt could not contract without generating tensile stresses because it was firmly held by cooler rock above or below which would not similarly contract. Tensile stresses acting towards the centres of the polygons fractured the rock. Later, the rock surface was polished and striated by a passing glacier.

8.3 See Fig. A8.3.

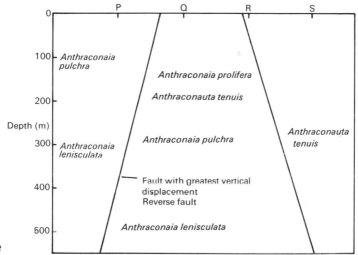

Fig. A8.3

8.4 (a) (i) See Fig. A8.4(a).

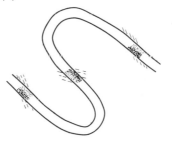

Fig. A8.4(a)

Graded bedding allows younging direction to be found. Where beds are the right way up, cleavage is usually steeper than bedding; where beds are inverted, bedding is usually steeper than cleavage.
(ii) Dolerite. Specimen 1, from the middle of the intrusion, has cooled relatively slowly so it has larger crystals than specimen 2, which has come from the chilled margin.
(b) The date for the gneiss gives the time when it was metamorphosed. During this metamorphism, low-melting point minerals melted to form liquid which crystallized to form pegmatite as the rocks cooled.

(c) The younger beds are overlapping the older beds as shown in Fig. A8.4(c).

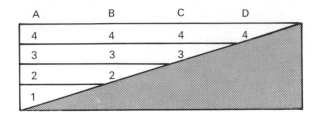

Fig. A8.4(c)

(d) B, D, H

(e) I: very cold;   II: cool;   III: warm, dry;   IV: warmer than III, wet; V: warm, dry;   VI: cool, wet.

8.5 (a) Intrusion J cuts the conglomerate so it is younger than the conglomerate. Intrusion I is cut by dyke D which is overlain by the conglomerate. The dyke is older than the conglomerate and intrusion I is older than the dyke. Intrusion J is younger than intrusion I; they cannot be part of the same large intrusion.

(b) An off-shoot of igneous rock K crosses and lies under the conglomerate. Lava flows do not show transgressive behaviour of this type.

(c) Dyke D is older than the conglomerate. Fault F is younger than the conglomerate.

(d) On both sides of the fault, the widths of the outcrop between the conglomerate and the greywacke are equal. Vertical movement on fault F would have made one outcrop wider than the other.

(e) Intrusion J cuts a dyke (top right) which cuts fault F.

8.6 (a) From old to young the faults are J, L, H and K.

(b) Between the Ordovician and Devonian Periods. Fault M cuts Ordovician rocks; M is cut by fault L which is pre-Devonian. Note that fault M could be late Ordovician.

(c) B

(d) (i) In an area between A and B about 80 m west of B. Here, there is a high rate of heat flow indicative of a cooling magma at depth. Hydrothermal deposits may be associated with this heat source.

(ii) The granite is younger than fault J. The heat flow measurements show that the granite is relatively young because it is still cooling. Fault J is pre-Devonian. The granite is probably younger than fault H. Fault H cannot be accurately dated. The heat flow graph seems to be unaffected by any faulting of the granite.

(e) C, G

(f) Precambrian: Original rocks of unknown type intruded by dolerite. Following metamorphism, a second set of dolerite dykes was intruded at right angles to the foliation of the gneiss.

Ordovician: Shales, sandstones and siltstones folded into a series of east–west trending anticlines and synclines. Faulting on J and M (same fault?) brought Precambrian and Ordovician into juxtaposition. Faulting then took place on L.

Devonian: Unconformable on Ordovician and on granite which may be of Silurian age.

Carboniferous: Unconformable with slight discordance on Devonian. Strongly discordant on Precambrian.

Ore deposits—probably of hydrothermal origin—formed in north. Faulting on K post-dates both the ore deposits and faulting on H. (Fault H is post-Devonian and pre-fault K. Otherwise, its age relationships are unknown.) Still-warm granite and associated hydrothermal ore deposits intruded in the south.

8.7 (a) See Fig. A8.7.

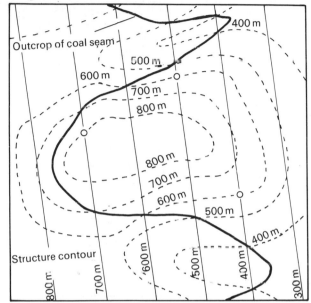

Fig. A8.7

(b) The seam dips at 44° to the ENE.
(c) See Fig. A8.7.

8.8 (a) The seam dips at 27° to the NW.
(b) See Fig. A8.8.
(c) Depth to seam CD: 1010 m; depth to EF: 898 m.
(d) See Fig. A.8.8.

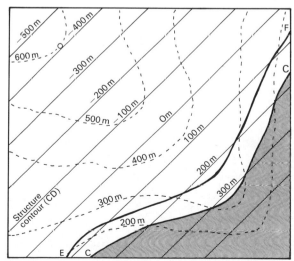

Fig. A8.8

8.9 (a) The coal seam dips at 16° to the SSE.
    (b) See Fig. A8.9.
    (c) Borehole A: 200; borehole B: 250 m.
    (d) Sandstone thickness: 190 m. The base of the sandstone is 200 m vertically below the top. The dip is 16° so the thickness = cos 16° × 200 m.

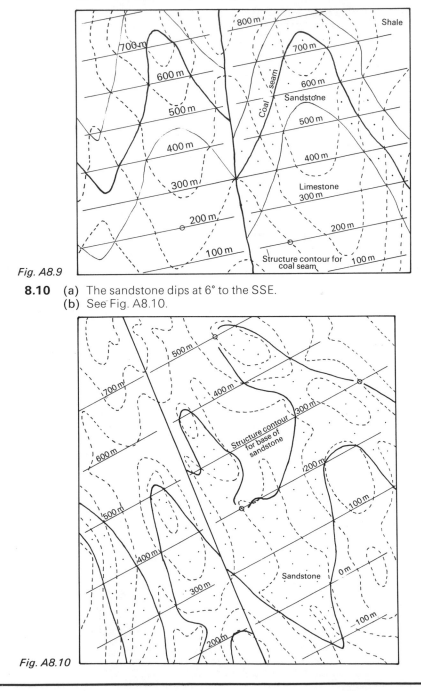

Fig. A8.9

8.10 (a) The sandstone dips at 6° to the SSE.
    (b) See Fig. A8.10.

Fig. A8.10

(c) The outcrop of the fault plane does not deviate as it crosses the topographic contours. Only a vertical structure would behave like this.

8.11 (a) Seam dips to SSW at 27°.

(b) See Fig. A8.11.

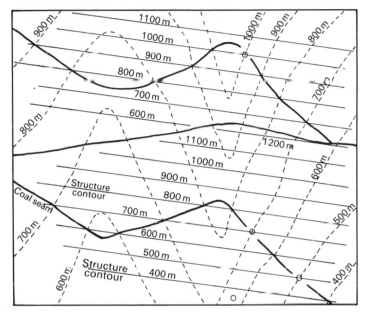

Fig. A8.11

(c) The rocks are thrown down 600 m to the N. (The 1200 m structure contour on the S side of the fault coincides with the 600 m structure contour on the N side of the fault.)

(d) The fault plane is not vertical; it dips steeply to the S.

8.12 The prominent rounded ridge consists of anticlinally-folded sedimentary rocks. In the middle distance the core of the anticline has been exposed by erosion. Just left of centre, the ridge is traversed by a deep chasm perhaps cut by a superimposed stream or eroded along the line of a fault. The valley on the far side of the ridge probably lies in a syncline. Near the top centre of the photograph, the beds dip gently to the right before steepening as they enter the distant part of the prominent anticline.

To the right of the ridge the rocks have been eroded down to form an irregular, hummocky landscape crossed by dry stream beds and gullies. The river emerges from a winding canyon on the extreme right of centre. The shapes of the islands and bars indicate that the river flows from top to bottom. Bottom right are alluvial terraces deposited either when the river flowed at a higher level or when the river was larger and was carrying much more sediment.

8.13 See Fig. A8.13.

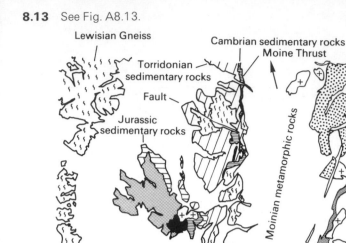

Fig. A8.13

8.14 (a) Five plates.
  (b) and (c) See Fig. A8.14.
  (d) Youngest crust: along the crests of the oceanic ridges.
   High heat flow: along oceanic ridge crests, island arcs and young Andean mountains.
   Strong negative gravity anomaly: along oceanic trenches.
   Explosive volcanic activity: at island arcs and young Andean mountains.
   Effusive volcanic activity: along oceanic ridges.
   Pillow lavas: along oceanic ridges.
  (e) A and A': moving away from each other at 3 cm yr^{-1}.
   B and B': moving past each other at 3 cm yr^{-1}.
   C and C': moving past each other at 3.3 cm yr^{-1}.
   D and D': since both lie within the same plate there is no relative movement between these points.
   E and E': moving towards each other at 4.3 cm yr^{-1}.
   F and F': no relative movement between these points.
   G and G': moving towards each other at 3.3 cm yr^{-1}.

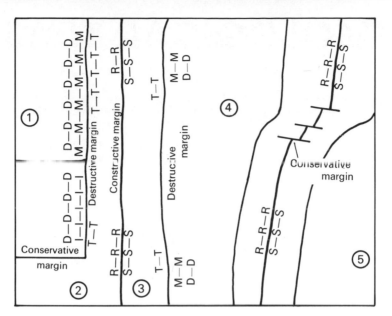

Fig. A8.14

8.15 (a) Topographic fit of coastlines. Mountains with pillow lavas, areas of Carboniferous glaciation and Permian desert sandstone suggest original continuity. Carboniferous and Devonian fossils are of organisms which may have lived on a single continent.

(b) Mountains may mark the position of an ancient destructive margin. Pillow lavas and trilobites indicate the previous existence of an ocean removed during plate collision.

9.1 (a) Delta volume: $5632 \times 10^9 \, m^3$
 (b) Age of delta: 53 640 years
 (c) He concluded this from the relationship between the river sediments and the rocks bounding the river valley. The sediments lie on rocks which are much older, so the Mississippi Delta must be geologically young.

9.2 (a) Kelvin had few details on the melting points of rocks and on thermal conductivities at high temperature and pressure. He could not accurately determine the rate of cooling.
 (b) Most of the Earth's heat is produced by the decay of radioisotopes. This heat has slowed the overall cooling of the Earth. Kelvin did not know about radioactivity.

9.3 (a) The half-life is the time in which the amount of a radioactive isotope or nuclide decays to half its original value.
 (b) (i) The ratio of parent to daughter atoms is 1:12. Since every daughter atom came from a parent atom, the ratio of surviving parent atoms to original parent atoms is 1:13. The number of half-lives which have passed is:

$$n = \frac{\log\left(\frac{1}{13}\right)}{\log\left(\frac{1}{2}\right)} = 3.7$$

The age of the sample is $3.7 \times 5 \, Ma = 18.5 \, Ma$.

 (ii) The ratio of parent to daughter atoms is 1:0.4, so the ratio of surviving parent atoms to original parent atoms is 1:1.4. The number of half-lives is:

$$n = \frac{\log\left(\frac{1}{1.4}\right)}{\log\left(\frac{1}{2}\right)} = 0.49$$

The age of the sample is 2184 Ma.

 (iii) The ratio of

$$\frac{^{238}U}{^{206}Pb} = 0.5405 = 1:1.85.$$

The ratio of surviving parent atoms to original parent atoms is 1:2.85. The number of half-lives is:

$$n = \frac{\log\left(\frac{1}{2.85}\right)}{\log\left(\frac{1}{2}\right)} = 1.51$$

The age of the Earth would be: $4500 \times 1.51 = 6795 \, Ma$. The ratio of

$$\frac{^{235}U}{^{207}Pb} = 0.0046 = 1:217.39.$$

The ratio of surviving parent atoms to original parent atoms is 1:218.39. The number of half-lives is:

$$n = \frac{\log\left(\frac{1}{218.39}\right)}{\log\left(\frac{1}{2}\right)} = 7.77$$

The age of the Earth would be: $710 \times 7.77 = 5517 \, Ma$.

(iv) Not all of the lead isotopes may have come from uranium. Some of the ^{206}Pb and ^{207}Pb may have been present when the Earth was formed.

(v) The modified ratio of

$$\frac{^{238}U}{^{206}Pb} = 0.9524 = 1:1.05.$$

The ratio of surviving parent atoms to original atoms is $1:2.05$. The number of half-lives is:

$$n = \frac{\log\left(\frac{1}{2.05}\right)}{\log\left(\frac{1}{2}\right)} = 1.0356$$

The age of the Earth would be: $4500 \times 1.0356 = 4660$ Ma.
The modified ratio of

$$\frac{^{235}U}{^{207}Pb} = 0.011 = 1:90.91.$$

The ratio of surviving parent atoms to original atoms is $1:91.91$. The number of half-lives is:

$$n = \frac{\log\left(\frac{1}{91.91}\right)}{\log\left(\frac{1}{2}\right)} = 6.52$$

The age of the Earth would be: $710 \times 6.52 = 4629$ Ma.
In part (iii), the ages are markedly dissimilar (6795 and 5517 Ma). Using modified U–Pb ratios gave answers which are both close to 4600 Ma. This suggests that the modified ratios are close to being correct.

(c) Radiometric ages are not completely accurate. The ± number reflects the range of analytical error. The errors are quoted at the 95% confidence limits. This means that if the measurement was made 100 times, 95 of the answers would be in the age range shown (between 3760 and 3580 Ma).

(d) Half-lives may not be accurately known.
Daughter atoms may have been present in the sample when it was formed and the amount originally present may not be accurately known.
Daughter atoms may be lost. For example, in the K–Ar method, argon readily escapes from minerals such as orthoclase.
Parent atoms may be lost. For example, uranium may be moved by leaching. (Where uranium has been lost, lead isotope ratios may be used to date rocks.)
Metamorphism may readjust parent–daughter ratios.

9.4 (a) (i) 250 Ma
(ii) No. Climatic changes alter rates of chemical weathering. Plate movements have taken land masses into areas of cooler or warmer climate where rates of chemical weathering differed. The area of exposed sodium-containing rocks may have changed. Global changes in rainfall may have altered rates of chemical weathering and volumes of water carried by rivers. Areas of internal drainage systems may have changed. (You may well be able to think of additional reasons.)
(iii) Sodium does not remain permanently in the oceans; it is recycled (e.g. halite may be found in evaporite deposits).
(iv) A sodium in sedimentary rocks; B sodium in metamorphic rocks; C sodium in igneous rocks; D chemical weathering;

E transport by rivers;   F sodium in rain and wind-blown spray.
(Note that only about 30% of sodium in rivers comes from weathered
rocks; the rest comes from rain and wind-blown spray.)
- **(b)** The residence time of sodium is 250 Ma. Sodium has a long residence time
because it is not extracted by animals and plants and it does not enter non-
evaporite sedimentary minerals. Calcium is used by organisms (e.g. corals,
molluscs, crustaceans, calcareous algae). Silicon is extracted by diatoms,
radiolarians and sponges. Manganese is precipitated as nodules of
manganese dioxide. (Elements of low abundance in sea water have short
residence times. Can you see why?)

9.5 **(a)** From old to young the sequence is: Old Lizard Head Series (schists);
Hornblende Schists (metamorphosed basalts); Man o' War Gneiss
(intruded as a migmatite); intrusive serpentinite (altered peridotite);
olivine gabbro; gabbro; olivine dolerite dykes; Kennack Gneiss (altered
granite); and red granite.
- **(b)** If an old (e.g. Precambrian) rock is metamorphosed and recrystallized,
radiometric methods record the origin of the rock as the time at which it
recrystallized. The rocks of the Lizard Complex have been metamorphosed
and recrystallized during the Palaeozoic. The age of 450 Ma for some of the
hornblende schists may show that they were only partially recrystallized by
Palaeozoic metamorphism. Alternatively, the schists may not have been
remetamorphosed, but they may have lost argon. This would make them
appear to be younger than they are.

PALAEONTOLOGY

10.1 **(a)** An orthorhombic form of calcium carbonate.
- **(b)** In warm (25–30 °C), clear, well-oxygenated sea-water of normal salinity
to depths of about 100 m.
- **(c)** Mutualism: A symbiotic relationship beneficial to both organisms.
The algae produce oxygen and food materials such as sugar, amino acids
and glycerol. They also use carbon dioxide produced by the polyps.
- **(d)** 1 To ensure that their mutualistic algae receive maximum illumination for
photosynthesis.
   2 To allow them to resist the destructive effects of waves.
   3 To increase their light-gathering capacities.
   4 To increase photosynthetic productivity in the dim light.
   5 Algae supply food which provides energy to build the aragonite
skeleton.
- **(e)** Reduces dependence on one food source (e.g. coral can survive and
grow even if it cannot actively feed.)
- **(f)** Wave action; grazing by fish, starfish and sea urchins; boring by animals
such as molluscs, sponges and worms.
- **(g)** Lagoons often have carbonate muds in their deeper parts and carbonate
sands in their shallower, more turbulent parts. The sediment may consist
of faecal pellets, sands consisting of foraminiferans and fragments of
corals and calcareous algae and mud produced by attrition and by the
activities of boring organisms. During storms, coarse-grained material
may be washed into the lagoon.

The fore-reef deposits are formed by debris which has fallen from the reef and collected like scree at the base of the outer reef. The reef talus consists of boulders grading out into sand and mud.

(h) Cambrian: stromatolites; archaeocyathids (cup-shaped, sponge-like organisms).

Ordovician: stromatolites; calcareous algae; stromatoporoids; tabulate corals; bryozoans.

Silurian and Devonian: stromatolites; calcareous algae; stromatoporoids; tabulate corals; rugose corals; bryozoans.

Carboniferous: stromatolites; green algae; calcareous algae; rugose corals; bryozoans; crinoids.

Permian: stromatolites; green algae; calcareous algae; sponges; foraminiferans; brachiopods (cone-shaped, rooted forms); bryozoans.

Triassic: calcareous algae; sponges; bryozoans.

Jurassic: calcareous algae; stromatoporoids; scleractinian corals; hydrozoans (resemble corals); bryozoans.

Cretaceous: calcareous algae; foraminiferans; scleractinian corals; hydrozoans; bivalves (dustbin-like rudists).

Cainozoic: calcareous algae; foraminiferans; scleractinian corals; hydrozoans.

10.2 (a) A species is a group of organisms which can interbreed in the wild and produce fertile offspring. Only about one-sixth of living animal species are proved species on this criterion. Morphological differences usually allow species to be established though there have been cases of males and females of the same species being classified as members of different species. Other differences (e.g. biochemical, physiological, behavioural) may be used as an aid to classification.

(b) Differences in morphology are all that can be used. This means that the classification of fossils is less certain than the classification of living organisms since more reliance is placed on the judgement of the taxonomist. It has been said that a 'good species' is what a 'good taxonomist' considers to be a 'good species'.

(c) See Fig. A10.2(c).

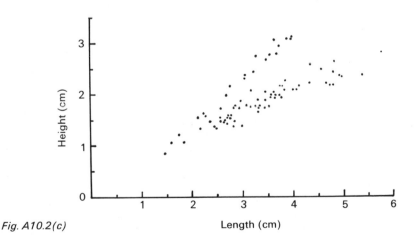

Fig. A10.2(c)

Two. In Table 10.2A, shells 1–60 are mussels (*Mytilus*) and shells 61–79 are trough shells (*Spisula*).

(d) See Fig. A10.2 (d).

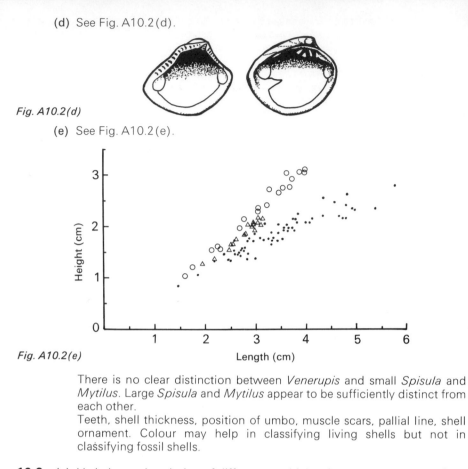

Fig. A10.2(d)

(e) See Fig. A10.2 (e).

Fig. A10.2(e)

There is no clear distinction between *Venerupis* and small *Spisula* and *Mytilus*. Large *Spisula* and *Mytilus* appear to be sufficiently distinct from each other.

Teeth, shell thickness, position of umbo, muscle scars, pallial line, shell ornament. Colour may help in classifying living shells but not in classifying fossil shells.

10.3 **(a)** Variation: a description of differences which exist among members of the same species.

Diversity: a description or measure of the variety of organisms. May be measured by the number of species in a family or order.

Maximum diversity occurred during the Devonian but brachiopods were also very diverse during the Ordovician, Silurian, Carboniferous and Permian.

(b) Inarticulate brachiopods have no teeth or sockets and the shell may be chitinophosphatic. The lophophore is unsupported. Articulate brachiopods have teeth and sockets and the shell is always calcareous. The lophophore may be supported. Lingulida.

(c) Extinction: 1 Increased competition from bivalves. 2 Evolution of improved predators. 3 Cooling of ocean by glaciation at end of Ordovician. 4 Removal of shelf seas when continents joined during the Permian. The sea also withdrew from remaining shelf areas because of the subsidence of spreading ridges. A great deal of salt was removed from the sea during the Permian so the oceans may have become brackish. Organisms which could not tolerate salinity changes may have become extinct. 5 During magnetic reversals, lethal radiation may have reached the Earth's surface.

There are many other possibilities.

Expansion: 1 Successful colonization of new habitats. 2 Evolution of new species better adapted to original habitats. 3 Increased availability and variety of food (e.g. by appearance of new forms of plankton). 4 Evolution of improved defensive mechanisms against predators. 5 Increased areas of shelf seas. 6 If a continent drifts off on its own, new species evolve in isolation. 7 General warming of seas. You may be able to think of other possibilities.

(d) Faunal province: a large marine area with a distinct fauna.

(e) Initially, the brachiopod faunas were distinct because they were on opposite sides of the wide Iapetus Ocean. Eventually, the ocean closed sufficiently for brachiopods from both provinces to colonize opposite shores of the ocean.

10.4 (a) Camerae: gas-filled chambers which provide buoyancy.
Septa: walls separating camerae. Provide strength. They buttress the wall of the shell against water pressure.
Suture line: the line along which a septum meets the inner wall of the shell.
Siphuncle: tube which passes through the septa and connects the camerae. The siphuncle adjusts the buoyancy of the shell by adding or removing fluid from the camerae.
Body chamber: the last, large shell chamber in which the animal lives.

(b) Ammonoid septa are fluted or corrugated. The suture lines are frilly or denticulate. Nautiloid septa are gently curved with smooth surfaces. The suture lines are simple curves. The corrugated ammonoid septa added strength so the relatively thin shell could withstand water pressure.

(c)
$$\text{Depth} = \frac{\text{pressure}}{\text{acceleration due to gravity} \times \text{density}}$$
$$= \frac{6 \times 10^{6}}{9.81 \times 1.025 \times 10^{3}} \, \text{m}$$
$$= \text{almost } 600 \, \text{m}$$

(d) Prevents siphuncle from bursting.

(e) They probably lived at depths of less than 600 m. Phylloceratids and lytoceratids had fairly strong siphuncles so they may have lived at depths of up to 450 m. Ammonitids had weak siphuncles so they may not have lived below 100 m.

(f) The centre of gravity would be well below the centre of buoyancy. When displaced, the test-tube would readily return to its original floating position. This position is very stable.

(g) Its centre of gravity is below its centre of buoyancy. This makes it stable so it does not spin easily.

(h) They would be poor swimmers. They would have difficulty staying the right way up when moving.

(i) 1 This would make the shell very strong allowing it to resist shell-crushing predators such as marine reptiles, bony fish, skates, rays, sharks, crabs and lobsters.
2 This would allow the ammonitid to swim and manoeuvre more rapidly so it could escape from predators.

10.5 (a) A structural, physiological or behavioural property which allows an organism to live in better harmony with its environment. An adaptation improves the organism's capacity to survive and, ultimately, to reproduce and leave descendants.

(b) *Echinocardium*: flattened shape with bilateral symmetry allows for easier movement through sand. Fragile test since protected by sand. Mouth moved to front for easier receipt of food. No Aristotle's lantern since food is not chewed. Anus moved to back for improved sanitation. The fascioles have ciliated spines which generate currents: the dorsal fasciole produces a current down the vertical funnel; the anal fasciole removes digestive waste; and the sub-anal fasciole creates a current along the sanitary tube or drain to the rear. There are spines adapted for protecting ambulacra, burrow building, burrowing, moving and current production. There are also tube-feet adapted for building the vertical funnel, gas exchange, feeding and sanitary drain building.

Micraster: the structural similarities are so strong that *Micraster* probably lived like *Echinocardium*. In *Micraster*, the anus is posterior and the lantern-free mouth has moved forward. The anterior ambulacrum lies in a groove; the remaining petaloid ambulacra probably carried respiratory tube feet. Granules on the petaloid ambulacra may have carried current-generating cilia. The presence of a sub-anal fasciole indicates that *Micraster* could create currents flowing away from the anus into a sanitary drain. The large plastronal tubercles suggest the presence of locomotory and digging spines similar to those on the plastron of *Echinocardium*.

In general, where there are no living forms which resemble extinct forms, we can only guess at the functions of structures found on fossils. Similar structures on living and fossil forms may well have had similar functions but we cannot be sure that this is so.

(c) Mussel: attached by strong byssus threads to sea floor. Since it does not move, the foot, the anterior part of the shell and the anterior adductor muscle are reduced. The strong shell provides protection against abrasion and predators.

Oyster: cemented to a hard surface by its left valve. Since the shell is completely immobile, the foot is rudimentary. The large posterior muscle lies centrally. There is no anterior muscle.

Scallop: attached by byssus when young but later becomes free-lying. Swims by rapidly opening and closing its valves. It rests on its right valve and it has a single large posterior adductor.

Cockle: lives in a shallow burrow with its posterior end often exposed. It has a strong globular shell not well adapted for rapid burrowing. It can use its large curved foot to allow it to hop over the surface of the sand.

Razor-shell: burrows rapidly by means of its large foot. It has short siphons so it feeds while lying at the top of the burrow. Since it is well protected by sand the shell is thin and open at both ends.

Gaper: remains stationary at the bottom of a deep burrow. The foot is small so it cannot burrow quickly. The siphons are so large they cannot be fully retracted. The shell is thin and open at both ends.

Piddock: bores mechanically into rocks. The foot forms a sucker which grips the end of the burrow. It has long siphons and the thin shell is open at both ends.

In general, bivalves show a variety of shell form, musculature, methods of attachment, and foot and siphon development which allows them to occupy a wide range of ecological niches.

10.6 (a) Dead shells will be transported towards the south-east. Waves approaching the shore from the south-west generate longshore currents which flow to the south-east.

(b) At position 2, the dog whelks are living on an exposed part of the shore. Their shells are relatively round, thick and heavy (52% have a height/width ratio of 1.5–1.59; 57% have a shell thickness of 0.20 cm or more; and 55% weigh 2.0 g or more). These characteristics probably allow the whelks to resist their high-energy environment. The shells have apparently been smoothed by abrasion.

The whelks in the harbour (position 1) are not genetically different from those at position 2. The shells are small, elongate, thin and light (54% have a height/width ratio of 1.6–1.69; 82% have a shell thickness of 0.10–0.19 cm; and 86% weigh less than 2.0 g). The shells are also ornamented with delicate ridges. These characteristics probably reflect the sheltered environment. The more restricted food supply in the harbour may also have contributed to the differences between the shells at positions 1 and 2.

The shells at position 3 are larger, thicker and heavier than those at position 2. (At position 3, 76% have a shell thickness of 0.20 cm or more and 86% weigh 2.0 g or more.) The shells at position 3 are also rounder than those at position 2. (At position 3, 82% have height/width ratios less than 1.6 while at position 2, 71% of the shells have similar ratios.) The shells at position 3 seem to be a transported fraction of a population similar to that at position 2. Small, light shells may have been carried further along the shore and thin-walled shells may have been broken to leave behind the shells found at position 3.

The shells in the raised beach deposits (position 4) resemble those at position 3 in being large, thick and heavy. (All have a thickness of 0.20 cm or more and 65% weigh 3 g or more.) They are also slightly rounder than the shells at position 2. The shells at position 4 would appear to represent a very small transported and preserved fraction of an original population which may have resembled that at position 2. Their study characteristics probably allowed them to resist transport perhaps over a long period. Smaller, thinner shells may have been broken or transported elsewhere.

(c) Probably because large, thick, heavy shells survive transport better than small, thin, light shells. Alternatively, the shells have been sorted into different fractions during transport. Collection of dead shells just happened to be where the larger, thicker, heavier shells had accumulated.

(d) The shells at positions 3 and 4 are relatively homogeneous groups sorted out from their original populations by transport. Thus, they show a smaller degree of continuous variation.

(e) The low height/width ratio of the shells at positions 3 and 4 is probably a result of their having thicker shells than the whelks at positions 1 and 2. As in the answer to (c), thick shells are well able to survive transport; or sorting has produced restricted accumulations of thick shelled whelks.

(f) It seems that the shells are only rarely preserved.

(g) The sand consists almost entirely of shell fragments.

(h) Not necessarily. The small sample of shells from the raised beach is probably not representative of the population from which it was derived. Positive conclusions about the environment cannot be drawn on the basis of such a small sample.

(i) No. Few whelks occupy environments as sheltered as that in the harbour. A sample in which such atypical shells made up 41% of the total would not be representative of the total shoreline population.

(j) The thick, heavy shells have probably been washed in from the harbour walls.

(k) Numerous species—especially those without hard parts—would not be preserved. It has been estimated that in a coral reef environment only about 2% of species would be preserved, while in a river bank environment only about 0.1% of species would be preserved. Even among organisms with hard parts, there may be differential preservation (e.g. organisms with chitinous skeletons may be better preserved than organisms with calcareous skeletons). Preservation also depends partly on habitat (e.g. benthonic organisms may be better represented in fossil assemblages than planktonic or swimming organisms). Small organisms may be better preserved than large organisms (e.g. large fish may be torn apart by scavengers while small organisms may be quickly buried). Thus, ecological relationships in the original living community can be only imperfectly established.

(l) Very small numbers of the original population may be preserved. The preserved organisms may have group properties unrepresentative of the original population. Organisms may have properties (e.g. thick shells) which improve their chances of preservation. Individuals of different ages may not be equally well preserved (e.g. adults may be better preserved than juveniles).

(m) Life assemblage: Fossils preserved in the place where the original organisms lived. The fossils are well preserved, entire and undamaged by transport. They show a complete range of different ages and sizes. The fossils represent part of an original living community.

Death assemblage: Fossils brought together by wind or water currents. The fossils show the effects of transport (wear, breakage, sorting, disarticulation, orientation and the fossil size is related to the sediment particle size). The fossils may have come from mixed environments so they are not representative of an original community.

The dog whelks in the raised beach deposits form part of a death assemblage.

(n) The sample will probably not be representative of the living population from which it was derived. The characteristics of the original population cannot be found by study of a small sample of fossils.

10.7 The rock is a ripple-marked sandstone of Carboniferous age. On the bedding plane surface are the trails or tracks of an organism which may have been a gastropod.

EARTH HISTORY

11.1 (a) Time units: era, period. Time-stratigraphical units: series, stage, system. Rock units: bed, formation, group.

(b) Zone: a biostratigraphical unit. A small part of a sequence containing distinctive fossils which make the zone unique. A zone is named after a single, index fossil. A zone may be defined by 1 the complete stratigraphical range of one species; 2 the stratigraphical range over which a species is particularly abundant; 3 a distinctive assemblage of fossils without regard for their stratigraphical ranges; or 4 a combination of overlapping ranges of selected species.

A zone fossil should have a narrow time range and it should be widespread, abundant, independent of facies and easily recognized. Swimming and planktonic organisms make the best zone fossils.

(c) Facies fossils are those which are restricted to particular sedimentary environments. Since they are not widespread, they have limited use in zoning. Such fossils are often derived from bottom-living organisms which lived in environments such as coral reefs, deltas, beaches or lakes.

(d) The marine Devonian of south Britain is zoned by means of ammonoids. The continental Devonian in the north is zoned by means of freshwater fish. Correlating rocks deposited under such diverse environments is very difficult. The Jurassic is mostly marine and it contains widespread, frequently changing ammonite faunas which provide numerous zones with narrow stratigraphic ranges.

11.2 (a) These are annual growth rings.

(b) In times of heavy rainfall, lakes would fill and overflow. Lakes would interconnect and fish would spread from one lake to the other.
In times of low rainfall, lakes would shrink or dry up. Fish populations would be separated. When a lake dried up, all fish would die.

(c) Variations in temperature, oxygen, pH, availability of food, salinity, mineral nutrients, predators, wind. (Wind creates waves which may stir up sediment and increase the opacity of the water. This reduces light available for photosynthesis. In a stratified lake, wind may destroy the stratification so the bottom layer of deoxygenated water rises to the surface. When this happens, all fish may die.)

(d) One year.

(e) Carbonate was precipitated during the dry season. Silt was washed in during the wet season.

(f) Removal of CO_2 causes a reduction in the carbonic acid level of the lake. With the rise in pH, carbonate solubility falls and it is precipitated

(g) The carbon fragments are algal remains. With the end of the dry season, many algae would die as the water became cooler and more opaque as sediment was washed into the lake.

(h) In the upper, oxygenated layer.

(i) When the fish died, they would have sunk into the deoxygenated layer. Here, they would be below the level of wave action so they would not be physically disarticulated. In the deoxygenated layer, decomposition would be incomplete and there would be no scavengers to tear the dead fish apart.

(j) 100 years. If the fish sank into the mud the time would be less than this.

(k) Fish populations would fall. As the lake dried up, salinity would rise. The fish would be concentrated into a smaller volume of water so food resources would fall. Oxygen is less soluble in warm water than in cool water. Decomposition of dead fish would also reduce oxygen levels.

(l) The lakes in which rock types II and III were deposited were shallow and unstratified so dead fish may have been torn apart by scavengers or broken up by wave action. Also, fish populations may have been low. These lakes would be strongly affected by seasonal desiccation which would reduce or destroy populations. At times of heavy rainfall fish might move in from other lakes. Adverse conditions during dry seasons might prevent rapid growth of the new population.

(m) Coprolites are fossilized droppings. They contain remnants of undigested food (e.g. fish scales, shell fragments).

(n) See Fig. A11.2 overleaf.

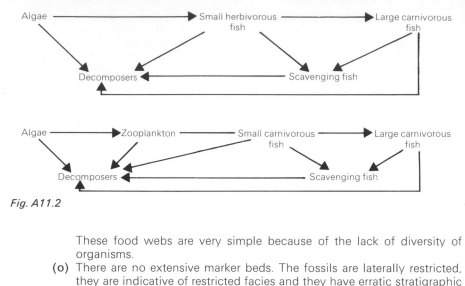

Fig. A11.2

These food webs are very simple because of the lack of diversity of organisms.

(o) There are no extensive marker beds. The fossils are laterally restricted, they are indicative of restricted facies and they have erratic stratigraphic distributions.

(p) Such a bed might contain a large number of species, some of which are not found in lower or higher horizons.

11.3 (a) One interpretation might be as follows.

   1 Low-energy environment. Probably in shallow lagoons or playa lakes subject to frequent desiccation. The sedimentary grains may have been blown into the lakes or lagoons from a dry, low-lying area.

   2 These are evaporites perhaps formed in intertidal salt marshes or on coastal flats bordering lagoons.

   3 The stromatolites and faecal pellets are indicative of a shallow marine or coastal lagoon environment.

   4 This is apparently a reef limestone.

   5 These appear to be fore-reef deposits formed by material falling down the front of the reef.

   6 These sediments seem to have been deposited in a marine basin receiving little terrigenous material. The sandstones may have been deposited by turbidity currents running from the north.

(b) A reef built by calcareous algae, bryozoans, sponges, brachiopods and foraminiferans separated a marine basin in the south from lagoons, salt flats and a low-lying land-mass in the north. In the north, the sediment is entirely terrigenous. Evaporites formed on the edge of the land while the lagoonal sediments have come mostly from the reef as a result of wave and biological erosion. Break up of the reef on the seaward side has produced coarse-grained fore-reef talus deposits. The marine basin received a little fine-grained sediment from the land and some turbidite sands, perhaps from behind the reef.

11.4 (a) Ophiolites are associations of rocks such as serpentinite, chert, pillow lava, peridotite, gabbro and basic dykes. They are thought to represent segments of oceanic crust and mantle pushed up or obducted at subduction zones onto island arcs or continental margins.

(b) In south Ayrshire, the early Ordovician rocks are ophiolites consisting of rocks such as radiolarian cherts, pillow lavas, serpentinites and gabbros.

Glaucophane schists are also present. In the same general area, sediments thicken very rapidly to the south. They may have been deposited on the landward side of an oceanic trench.

The Lake District volcanic rocks (andesites, rhyolites, tuffs, ignimbrites and agglomerates) are of the type which could well have formed in an island arc environment. They also have chemical similarities to volcanic rocks presently forming in island arcs.

There is no evidence of a constructive margin.

(c) Deposition of thick greywackes and mudstones took place on a rapidly subsiding sea floor. Large scale volcanic activity produced andesites, rhyolites, ignimbrites, pillow lavas and tuffs. The volcanic rocks are of types formed well behind subduction zones. Submarine fans were built against fault scarps by slumps and turbidity currents triggered by fault movements.

(d) The main phase of mountain building took place during the early Ordovician. At this time the Moinian and Dalradian rocks were strongly metamorphosed and deformed. Granites were formed by partial melting during metamorphism. Away from the Scottish Highlands, deformation was later and much less severe with slate being the dominant metamorphic rock. During the Silurian Period the ocean had almost closed and movements may have begun on the Highland Boundary and Southern Upland Faults. The final phase of the Caledonian Orogeny occurred during the Devonian Period with continental collision and the intrusion of granites in the Highlands, Southern Uplands and Lake District. Subsequent to their final uplift, the Caledonian Mountains were deeply eroded.

11.5 (a) One interpretation might be as follows. The general sequence of conglomerate, sandstone, siltstone and shale suggests that deposition took place in rapidly subsiding basins (1 and 3) separated by a slowly subsiding intermediate area (2). Area 2 may have been separated by faults from the basins on each side. The supply of terrigenous sediment seems to have ceased as the basins filled and the limestone may have been deposited in shallow water.

The sequence at 2 is condensed.

(b) (i) One interpretation might be as follows. The presence of limestone at the top and bottom of the sequence might suggest that sedimentation is cyclic and that incomplete and interrupted cyclothems are shown here. If so, the sequences may have been deposited in a subsiding basin into which deltas advanced. The lower limestone would represent off-shore, marine conditions. The shale would represent terrigenous sediment deposited in front of the advancing delta. After depositing the lower tongue of sandstone the delta may have built a second, upper lobe in a different direction. The delta lobes were carried beneath the sea by continued subsidence and the cycle began again.

(ii) No. Deltas may have a great variety of shapes and depositional histories. Beds will not necessarily thicken towards an adjacent landmass.

(c) See Fig. A11.5 overleaf.

The symmetry and thickness variations suggest that the deposit is an alluvial fan or cone deposited by a mountain stream dumping its load where it ran onto a plain. The deposit may also be a submarine fan or cone deposited at the mouth of a submarine canyon or submarine channel.

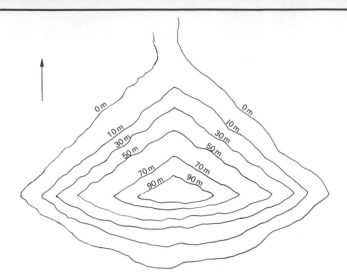

Fig. A11.5

11.6 (a) Brockram is Permian breccia in which large fragments of older rocks (particularly Carboniferous Limestone) are embedded in a red sandy matrix. The breccias were formed as screes and alluvial fans on the sides of raised fault blocks (horsts). Sometimes, the screes filled valleys which were probably wadis.

(b) Millet-seed sand is desert sand consisting of near-spherical, frosted quartz grains. Grain collisions in air are uncushioned so grains quickly lose projections and become well-rounded. The effects of attrition are enhanced by rapid movements produced by strong winds and by the fact that the grains roll and saltate over large distances.

(c) From the orientations of dune-bedding foresets. In a transverse aeolian dune, the steep leeward slope forms at right angles to the wind direction. Since the cross-bedding foresets form parallel to the leeward slope, measurement of the true dip of the foresets allows the wind direction to be found.

(d) The salts precipitated as the Zechstein Sea evaporated. Since the K salts are more soluble than halite, a more extreme degree of evaporation is required before they will precipitate. Hence, K salts occupy a restricted area in the middle of the basin of evaporation because they will not precipitate until the water has almost gone.

11.7 (a) The sediments exhibit cycles of marine transgression and regression. For example, in the London Basin, the Woolwich Beds are marine in the east and they grade through estuarine deposits into fluvial Reading Beds in the west. The marine London Clay overlies the Woolwich and Reading Beds. In the west, the London Clay grades up into fluvial Bagshot Beds.

(b) Fossils: marine beds are characterized by fossils such as sharks' teeth, nautiloids, corals, *Nummulites*, brachiopods, barnacles and echinoderms. There are also numerous marine bivalves and gastropods. Note that marine beds may contain numerous fossils of land plants. Estuarine beds may have crustacean burrows like those found in modern shoreline sediments. Fluvial deposits contain freshwater bivalves and gastropods. Characteristics of sediments: for example, glauconite is found only in marine beds; shoreline beds deposited during transgressions may contain flint pebbles derived from the Chalk; and fluvial beds may contain strongly oriented cross-bedding indicating transport from the west.

(c) The climate was probably warm because about half of the plant fossils have modern representatives living in the tropics. Some of the plants have living representatives which are not found in the tropics. These may have lived in dry or cool, elevated areas. Animal fossils such as those of crocodile, turtle and hippopotamus are also indicative of a warm climate. Kaolinite deposits in Devon have apparently been produced by the lateritic weathering of granite. Such weathering is indicative of a hot, humid climate.

EARTH RESOURCES

12.1 (a) (i) Even small degrees of disturbance might separate grains and create small fractures which would greatly increase permeability.
(ii) To remove dissolved air which might form bubbles in the sample. The bubbles would reduce permeability.
(iii) To prevent removal of fine-grained material from the sample. This would increase permeability.
(iv) The difference in hydrostatic pressure between two points divided by the distance between them. $N\,m^{-3}$.
(v) Hydraulic gradient is measured entirely within the sample.
(b) (i) $m\,s^{-1}$ (ii) $k = 10^{-3}\,m\,s^{-1}$ (iii) $k = 10^{-1}\,m\,s^{-1}$
(c) (i) Rock should be impermeable.
(ii) Should be capped with impermeable material so rainfall will not reach waste. This will prevent contamination of water supplies with toxic liquids or leachates from solid waste.
(d) (i) The floor of the site should have been sealed with clay.
(ii) It has not been thoroughly compacted so large voids are present which allow water to flow through.
(iii) 320 000 years.
(iv) The bottom clay is not sufficiently extensive so toxic materials could reach limestone by travelling above the clay.
(v) From A to B: about 560 years.
From A to C through siltstone: about 230 years.
From A to C through siltstone, sand and siltstone: about 55 years.
The time depends on the direction of groundwater movement. It may not move from A towards B and C.
(vi) See Fig. A12.1.

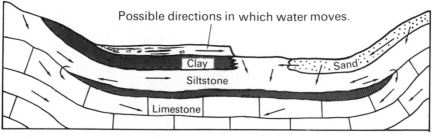

Fig. A12.1

12.2 1 diamonds in kimberlite; 2 cumulates; 3 liquid immiscibility; 4 pegmatites; 5 Sn–Cu–W of Cornwall; 6 Pb–Zn of south Pennines; 7 metamorphic processes; 8 contact metasomatism; 9 mechanical accumulation; 10 banded iron formations; 11 residual processes; 12 secondary or supergene enrichment; 13 exhaled hydrothermal solutions.

12.3 (a) (i) A geochemical anomaly is a concentration of elements in materials such as soil, water, sediment or plants which differs significantly from normal, background values.
Ore bodies represent anomalously high concentrations of particular elements. Elements moved for some distance by the operation of processes such as weathering and erosion may still show concentrations well above normal.

(ii) See Fig. A12.3.

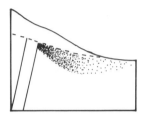

Dispersion by ground water

Fig. A12.3A

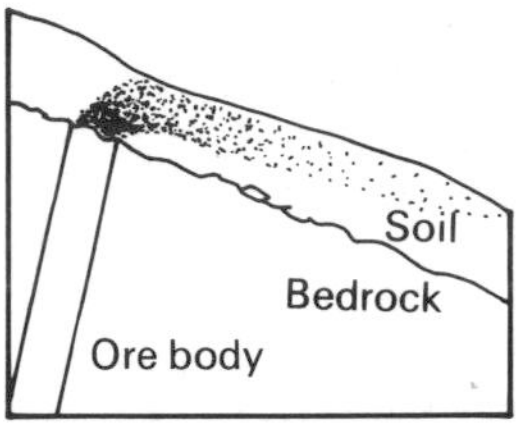

Dispersion by weathering
and soil creep

Fig. A12.3B

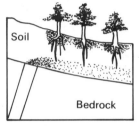

Dispersion by plants

Fig. A12.3C

(iii) Before glaciation, the Afon Wen flowed along the now-buried river channel so copper-rich sediment was carried to the west of the ore bodies. Subsequent glaciation and the activity of post-glacial streams smeared out the geochemical anomaly.

(b) (i) Would not be detectable because gold is non magnetic.

(ii) Ilmenite is strongly magnetic. The deposit would be detectable if the ilmenite was sufficiently concentrated.

(iii) Not detectable; fluorite is non magnetic.

(iv) The diorite intrusion would be detectable because the strength of its magnetism contrasts strongly with that of the limestone. However, the presence of chalcopyrite could not be detected.

(v) Could be detected if sufficient haematite was present.

(vi) Not detectable. There would be little contrast in the strength of magnetism between the vein and the granite especially if iron-containing minerals were present in the vein.

12.4 (a) (i) The amount of heat given off by the complete combustion of a given mass of the fuel. The SI unit is J kg^{-1}.

(ii) The process of diagenesis and metamorphism by which coals are formed from peat.

(iii) The rank of a coal is a measure of the degree of coalification which has been reached. The main change with increased coalification is an increase in carbon content relative to moisture and volatiles. Rank is often expressed as the percentage of carbon in the coal.

Increase in rank results mainly from the effects of rising pressure and temperature as the coal is buried to progressively greater depths. Heat and pressure generated by folding, faulting and igneous intrusions also contribute. From wood to anthracite, the main changes are an increase in carbon accompanied by a decrease in oxygen. Hydrogen also decreases while nitrogen shows a less regular change.

(b) (i) Mostly lost as CO_2, H_2O and CH_4.

(ii) The nitrogen comes from the bacteria which partly decompose the wood. Bacteria may contain 13% nitrogen.

(iii) When coal burns, it combines with oxygen. The more oxygen with which it can combine, the more heat will be given off. Coals of low rank contain a high proportion of oxygen so they can combine with less oxygen than coals of high rank.

(c) (i) Fixed carbon is the carbon left when all the volatiles have been removed.
Volatiles such as hydrocarbons and phenols contain carbon.

(ii) 100 t of peat would give 74 t of lignite, 43 t of bituminous coal and 30 t of anthracite.

12.5 (a) Shales contain more radioactive elements such as uranium, thorium and potassium.

(b) (i) Porous rocks may hold water or petroleum. These fluids contain hydrogen.

(ii) Gypsum ($CaSO_4.2H_2O$) contains hydrogen. Anhydrite ($CaSO_4$) has no hydrogen.

(c) Good conductivity results from the presence of pore water which contains dissolved salts. Dry, non-porous rocks such as halite and anhydrite would have low conductivities. Porous rocks containing petroleum or fresh water would also be poor conductors.

(d) Sound wave velocity depends partly on the compressibility of the rock. Pore spaces contain gas or liquid which are relatively compressible. This reduces the velocity of the sound waves.

(e) Allows good correlation. Gives details of extent and properties of reservoir and of structures such as folds, faults and unconformities.

12.6 (a) (i) The discovery of new reserves more than compensated for petroleum which had been extracted.

(ii) Prior to 1960, North America had been quite well explored and most major oilfields had been found. High rates of production ate into reserves. Other areas were perhaps geologically less well known and more detailed and more sophisticated exploration found significant reserves, often in offshore fields.

(iii) World reserves in 1960: 40 788 million tonnes.
Total production in 1960: 1091 million tonnes.
Reserves would be exhausted after 37.39 years, i.e. in 1997.
World reserves in 1980: 89 346 million tonnes.
Total production in 1980: 3085 million tonnes.
Reserves would be exhausted after 28.96 years, i.e. in 2008.

(b) (i) The lifetime of a resource is the length of time for which it will last. The lifetime may be extended by the following: finding new reserves, e.g. by remote sensing; using better methods of extraction, e.g. enhanced oil recovery techniques; reduction in consumption, e.g. by price increases or economic recession; change of use, e.g. resource may be replaced by cheaper, more plentiful material; resource material may be used more efficiently, e.g. insulating buildings reduces the use of fuels; more efficient recycling, e.g. lead can be recovered from old car batteries; governments, cartels or multinational companies may control production, e.g. by introducing conservation measures.

(ii) Renewable resources (e.g. ground water, tidal power) are replaced over reasonably short intervals. Non-renewable resources (e.g. coal, oil) are not replaced once extracted.

(iii) Resources of high place value (e.g. gravel, building stone) are of low monetary value so it is not worthwhile to transport them over long distances. Resources of low place value (e.g. oil, diamonds) are of high monetary value so they are worth transporting over long distances.

(iv) See Fig. A.12.6.

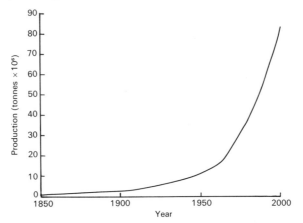

Fig. A12.6

Exponential growth: growth of the type shown by savings under compound interest, i.e. the amount of money added each year is a proportion of what was already there. Where resource production is increasing at r % per year, production after n years is:

$$\text{final production} = \frac{\text{initial}}{\text{production}} \left[1 + \frac{r}{100} \right]^n$$

Exponential growth can also be described by saying that the log of the amount of resources produced increases in a straight line with time.

From the graph, you should find the doubling times to be: from 1850, 35 years; from 1900, 23 years; and from 1950, 18 years. The formula giving the number of years taken to double production is:

$$\frac{\text{number of years}}{\text{to double production}} = \frac{\log 2}{\log \left[1 + \dfrac{r}{100} \right]}$$

$$\approx \frac{70}{r}$$

where production is increasing at r % per year. Where $r = 2$%, the doubling time is 35.0 years; where $r = 3$% the doubling time is 23.4 years; and where $r = 4$% the doubling time is 17.7 years.

Production before and after doubling: from 1850, 1 million to 2 million tonnes; from 1900, 2.692 million to 5.384 million tonnes; and from 1950, 11.800 million to 23.600 million tonnes. After 1950 doubling of production requires the finding and exploitation of large quantities of new reserves.

12.7 (a) An unconfined aquifer is not covered by an impermeable bed. A confined aquifer has an impermeable bed above and below.

(b) Loss of water from the land surface by evaporation from rivers, lakes, etc. plus transpiration from plants.

(c) During windy weather, tiny drops of sea water are carried into the atmosphere.

(d) Dykes restrict the movement of groundwater.

(e) Peat intercepts rainfall and holds it like a sponge, so much of it does not reach the aquifer.

(f) See Fig. A12.7.

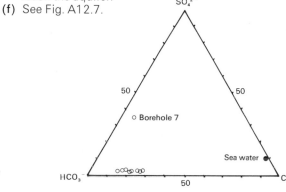

Fig. A12.7

(g) No. On the triangular diagram the sea water and the water for borehole 7 plot in completely different places. The water in borehole 7 has been produced by the mixing of shallow, fresh groundwater with deeper mineralized water rich in calcium sulphate. Ingress of the mineralized water is caused by over pumping.

(h) The lochs have small volumes. Since they are shallow, waves will stir up mud and make the water turbid. Shallow lakes or lochs may also be rich in decaying organic matter. The water may also be polluted by farm animals and by roosting sea birds.

(i) Borehole 1: 17 280 litres. Borehole 9: 1 036 800 litres.

(j) Fall of the water table in a well because discharge exceeds recharge.

(k) Rapid drawdown: the water can be replenished only very slowly because the aquifer has a low permeability or because rainfall has been low.

Slow drawdown: the aquifer is fully charged and permeable.

(l) A local, cone-shaped lowering of the water table around a well because water has been pumped out more quickly than it can enter the well from the aquifer. (Also called a cone of exhaustion.)

(m) The low Eh values in boreholes 7 and 9 indicate that the water is not saturated with oxygen.

12.8 (a) (i) Sapropel ('putrid mud') is black mud produced by the partial anaerobic decomposition of organic remains.

In restricted basins and on continental slopes there is abundant phytoplankton and deposition can take place under low-energy, anaerobic conditions.

(ii) Kerogen is organically-derived matter disseminated through a source rock. It also describes bituminous matter in oil shale.

(iii) Where inorganic sediment is being rapidly deposited, organic material would not attain sufficiently high concentrations to form source beds. In areas of slow deposition, much organic matter would be eaten by scavengers before it could be buried.

(iv) Carbon—74% increase; hydrogen—59% increase; oxygen—96% decrease; sulphur—no change; nitrogen—96% decrease.

(v) There is an overall loss of material in forming crude oil from phytoplankton. Carbon and hydrogen are lost less quickly than the average rate of loss, so they show a proportional increase in crude oil. To remain at the same percentage, sulphur is lost in the same proportion as the total loss.

(vi) Hydrocarbons.

(b) (i) See Table A12.8.   (ii) See Fig. A12.8(b).

Geothermal gradient ($^\circ C\,km^{-1}$)	Depth at which temperature of 75°C reached (km)	Depth at which temperature of 170°C reached (km)
10	7.5	17.0
20	3.75	8.5
30	2.5	5.67
40	1.88	4.25
50	1.5	3.4
60	1.25	2.8

Table A12.8

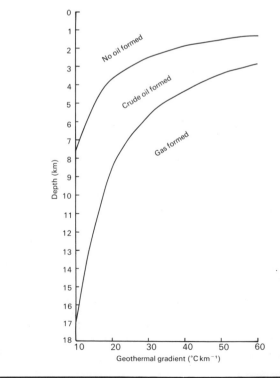

Fig. A12.8(b)

(iii) 3.0–6.8 km   (iv) 15° km^{-1}; 34° km^{-1}

(c) (i) See Fig. A12.8(c).

(ii) 48–114 °C

(iii) There is no petroleum. At this temperature the oil has been destroyed and the gas has escaped.

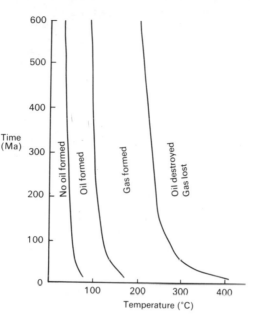

Fig. A12.8(c)

12.9 (a) (i) Granites. The granites are rich in heat-producing radioactive elements such as uranium, potassium and thorium.
Mesozoic sedimentary basins. The basement rocks under the basins may be slightly hotter than those in other areas. The heat is carried upwards by convection of ground water.

(ii) Palaeozoic sedimentary rocks tend to be hard and compact with low porosities and permeabilities. At depths where water temperatures are suitable for geothermal development, the rocks tend to form poor aquifers.

(iii) 1.48×10^{14} J. United Kingdom electricity generation is about 6750 times greater.

(b) (i) Geothermal gradient is the difference in temperature between two points on a vertical line within the Earth, divided by the difference in depth between them.

(ii) Thermal conductivity is a measure of the rate at which heat flows through a rock. The rate of heat flow between two points is proportional to the temperature difference between the points.

(iii) 9.5 km; 35.2 °C km^{-1}.

(iv) The geothermal gradient will be reduced. High thermal conductivity smooths out large temperature differences.

(v) It will be improved. The covering rocks provide good insulation and so keep the granite at a high temperature.

(vi) The thick layer of insulating sedimentary rocks traps heat escaping from the basement. Since the rocks at the base of the sequence remain hot while those at the top are cool, there is a high geothermal gradient.

(c) Granite is hot and dry, so cold water is pumped down one borehole to be heated as it passes through fissures enlarged by hydraulic fracturing before coming back to the surface through a second borehole. Hot, salty ground water is extracted from sedimentary rocks. After cooling, it is reinjected to maintain reservoir pressure.

Steam can be used to drive generators to produce electricity while hot water can be used for space heating. In the United Kingdom, granites have the potential to produce electricity equivalent to ten years supply. However, there are considerable technical difficulties in extracting steam from a depth of 6 km in granite at the required temperature of 200 °C. Also, you will see from (a)(iii) that it would take a very large number of geothermal stations to make a significant contribution to generating capacity. In this country, the hot water from sedimentary rocks is usually at a temperature of about 60 °C. Work on this limited resource has been largely curtailed.

12.10 (a) (i) Shear strength $= 151 \, \text{kN m}^{-2}$
(ii) Shear strength $= 270 \, \text{kN m}^{-2}$

(b) (i) Shear strength decreases. Caused by reduction of both cohesion and the angle of friction.
(ii) Old cuttings will show signs of failure as they settle to lower angles.
(iii) The table suggests 16° or less for a cutting designed to last for 60 years. Study of natural slopes suggests that the Clay eventually settles to a slope angle of 11°.

(c) (i) Shear stress increases, normal stress decreases. Slip would be more likely on a steeply-inclined plane.
(ii) No. Hydrostatic pressure has no shear component.
(iii) A small amount of water in a granular material (e.g. soil) increases cohesion because forces of surface tension in the film of water between the grains tend to hold them together. A large amount of water reduces the shear strength of the soil, while very large amounts may make the soil behave as a liquid. Water may weaken rocks by removing soluble cement (e.g. from calcareous sandstone) and by increasing the rate of weathering. Water may also lubricate planes of weakness such as bedding planes and joints. The increased growth of vegetation in damp conditions improves slope stability because roots provide a mesh which holds soil in place.

(d) 1 In the dry sand, the slope is maintained by internal friction. The moist sand has some additional cohesion provided by the small amount of water between the grains.
2 The large amount of water in the wet clay has acted as a lubricant and so has reduced forces of internal friction.
3 The crushed sandstone consists of angular fragments which lock together better than the grains of dry sand. The crushed sandstone has a greater degree of internal friction.
4 Small grains act as wedges between large grains and so resist their tendency to move over each other. For this reason the sand of mixed grain size has more internal friction.

(e) (i) In the loose sand, shear resistance (shear strength) rises with movement. As the grains move into more stable positions the volume of the sand decreases. Eventually, the volume and shear strength reach constant values. In the compacted sand, the shear resistance increases more rapidly with movement. After reaching a peak of shear strength the shear resistance decreases. During movement, the tightly-packed grains move up and over each other, so the volume of the compacted sand increases.

(ii) Both sands attain the same degrees of compaction with equal porosities. Since the physical condition of both sands is the same, they respond similarly to shear stress.

(f) (i) Cutting: vegetation is removed and parts of the hillslope are steepened. Unloading clay causes expansion so tension-cracks open, causing an increase in permeability, softening and a reduction in strength. Pore pressures are reduced for several months.

Filling: the slope is steepened and a plane of weakness is created between the fill and the underlying hillside. On the filled slope, there is an increase in height and weight. There may be a rise in pore water pressure so shear strength is reduced.

Both processes produce a general reduction in slope stability.

(ii) Roads should be as narrow as possible and high cutbacks should be avoided.

(iii) To prevent entry of water and to increase strength. Plant material will decompose and create cavities and other weak areas in the embankment.

(iv) Ground motions may trigger debris avalanches, especially under wet conditions. Material flung aside by blasting may overload the slope and cause debris slides.

(v) Safe disposal sites (e.g. hollows) should be found.

(vi) Balanced fill construction reduces the amount of excavation and it eliminates the need to dispose of waste.

(vii) In steep terrain, full bench construction is most suitable. It conveys maximum slope stability and it causes least disturbance of the slope below the road. It has the disadvantage that cut material may have to be moved a long way to be disposed of in stable areas.

Sliver fill produces little disturbance to the downslope side of the road but the fill cannot be properly compacted.

Balanced fill construction is not suitable because it produces a great deal of disturbance on the downslope side.

(viii) Cross drains should be of adequate number and size. Water should not be routed onto unstable areas. Drains should extend beyond fill. Boulders at drain outlets will prevent gully erosion. On very gentle slopes the road can be sloped outwards.

(ix) The trees prevent gully erosion and their roots bind the soil and reduce the probability of landslides.

(x) The construction should avoid loading the head and undercutting the toe of the potential slump. It should shift the centre of gravity of the slump block to a lower, more stable position.

(xi) As the hillslope gradient increases, the width of the disturbed area increases exponentially. In very steep terrain, the width of the disturbed area is disproportionately large. See Fig. A12.10 overleaf.

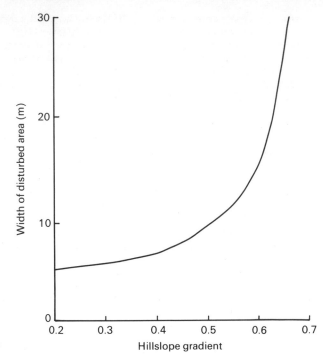

Fig. A12.10

12.11 (a) Rocks may slide from the east bank of the reservoir because water penetrating the limestone will lubricate the contact between the limestone and the shale.
Faults on the west bank may cause water loss or foundation instability.
Glacial sand and gravel are highly permeable; they would not provide a suitable foundation.
The reservoir would suffer severe water loss through the soluble limestone.

(b) Favourable factors: the river is fed by glaciers so high summer demands could be met. A syncline is a strong structure so the dam would be well-founded. The upstream dip of the beds reduces water loss and also provides a strong foundation.
Unfavourable factors: sediment supply in summer would cause rapid silting up. Water would be lost through joints. Since the site is located on a syncline, beds on the banks would dip towards the reservoir. Rocks may slide into the reservoir.

12.12 (a) Marl is a calcareous mudstone, cornstone is a concretionary limestone and cementstone is an argillaceous, dolomitic limestone. With the calcareous sandstone and shale, all rocks contain soluble carbonates which may be removed by slightly acidic ground water.

(b) When the water was lowered for site investigation, 0.7 million litres a day were still escaping. At this time all of the water lay on the topmost shale.

(c) The rocks immediately under the dam are cavernous.

Rock unit	Flow of water (litres per minute per metre)
Top sandstone	118.85
Top marl	4.48
Middle sandstone	21.87
Bottom marl	0.66
Bottom sandstone	0.33
Lava	0.17

(e) The removal of calcite cement has left loose quartz grains which have been washed away to leave cavities which have greatly increased the permeability of the rock.

(f) The permeability of a rock is related to the sizes and shapes of the intergranular spaces and their interconnections. Removal of $CaCO_3$ from marl would leave a fine-grained matrix with small intergranular spaces which would make the rock only slightly permeable. Removal of $CaCO_3$ cement from the sandstone would create large intergranular spaces which would markedly increase the permeability of the rock.

(g) See Fig. A12.12.

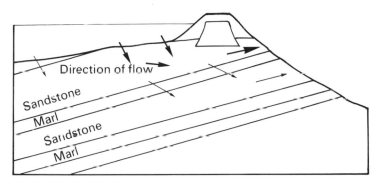

Fig. A12.12

(h) The weak grout mix would move easily into the small cavities. The following thicker mixes would push the weaker mixes further out into the surrounding rock and they would fill and seal the large cavities.

(i) Small volumes of thin mix would suffice to fill the small cavities. Large volumes of thick mix would be required to fill the large cavities.

(j) The cavities had been largely filled by the Stage 1 injection. The Stage 2 and 3 injections were required to seal the remaining small volume of unfilled cavities.

12.13 (a) Here, wire mesh has been spread over a near-vertical cutting to prevent falling rocks from reaching the road.

(b) These are stone-filled herringbone drains. Efficient drainage ensures that the slope does not retain water and become unstable during periods of heavy rainfall.

12.14 **(a)** (i) Subsidence is caused by collapse of the roof between pillars and by erosion and oxidation of the pillars. Voids migrate upwards during failure and crownholes appear at the surface.

(ii) The backfilled material in the old quarry will probably be poorly consolidated. Part of a building overlapping the quarry will subside while the rest of the building remains stable. The building may suffer severe damage.

(iii) Window and door frames are distorted so they may not open or shut. Lintels and walls show cracks. Walls deviate from vertical.

(iv) If layers of compressible peat and soil are in the drift, they will suffer volume reduction under the weight of buildings and subsidence will result.

(v) Subsidence results because the loss of pore water pressure causes compaction of the sediment. Venice and Mexico City have suffered subsidence because of water extraction from aquifers.

(vi) The house may settle by up to 10 cm as extraction of soil water lowers the water table and reduces pore pressure. If the tree is felled, the water table rises again. The building is pushed up (at rates up to $6 \, mm \, yr^{-1}$) and new cracks may appear in the walls.

(vii) Sediment may subside into sinkholes in the limestone. Such effects have taken place in the goldfields near Johannesburg, South Africa.

(b) (i) Tension occurs on the outer edge of the area of draw. Cracks appear in buildings and roads. Compression occurs on the inner side of the area of draw. Previously-formed tension cracks are closed up and new cracks appear. Paving slabs are pushed up.

(ii) Movement concentrated along pre-existing faults may produce narrow step-like surface features.

(iii) The area of draw increases.

(iv) Downward movements only. Damaging tensional and compressive forces are minimal. When the face is abandoned, the effects of draw are marked and a step-like feature may appear at the surface.

(v) Drilling may detect voids, pieces of pit prop and water-marked stones. Sealed mineshafts also produce detectable magnetic anomalies.

(c) (i) A: 1.24 m; B: 0.66 m; C: 1.32 m; D: 2.48 m.

(ii) Subsidence increases as the seam thickness and the width of the workings increase. Subsidence decreases as the depth to the seam increases.

(d) (i) Oil extraction caused lowering of pore fluid pressure. This allowed particles to pack more closely, so reducing the volume of the strata. Water injection restored pore pressure and increased the volume of the beds.

(ii) The land may sink below sea level. In Long Beach, dykes were built to prevent flooding.

12.15 **(a)** By evaporation of seawater and in fine sea-spray. Iodine can be carried in solution in water droplets, on dust particles and as a gas.

(b) 280 t; 93.3 m.
The figures would be increased. The greater the loss of iodine, the greater the increase.

(c) No. It would require the chemical weathering of huge quantities of rock in the relatively short time since the last Ice Age.
From rain, sea-spray, atmospheric dust and seaweed fertilizer.

(d) Concentration factor above igneous rocks = 18.6.
Concentration factor above sedimentary rocks = 2.53.
Iodine may be held in soil by being adsorbed by humus, clay and other silicates or it may be actively absorbed by microorganisms. Whatever the reason, soils derived from igneous rocks seem better able to retain their iodine.

(e) Coastal location means that Galway receives iodine from the sea. Galway has most igneous rocks, and soils above igneous rocks are relatively rich in iodine. The daily requirement of about 200 μg of iodine can readily be obtained from food. It is difficult to see why Galway potatoes have relatively low levels of iodine. Perhaps different strains of potatoes differ in their ability to absorb iodine.

(f) Glaciation removed soils and meltwaters dissolved and removed iodine. In the short time since glaciation, soils have not had time to build up high concentrations of iodine.

12.16 (a) Granite and olivine peridotite: the peridotite shows much less reflectance than the granite, showing it to be much darker. In the peridotite, the strong absorption band at about 1 μm is due to the presence of Fe^{2+}. The granite seems to be lacking in iron.
Fossiliferous limestone and red sandstone: the limestone has a higher reflectance than the sandstone, showing it to be paler. The sandstone shows well-developed water absorption bands at 1.4, 1.9 and 2.2 μm. There are no well-defined absorption bands for iron. The limestone shows a strong CO_3^{2-} absorption band at 2.3 μm. The weak absorption band at 1.9 μm may be due to the presence of CO_3^{2-} or water. There is also a very weak water absorption band at 1.4 μm.
Serpentine marble and hornblende schist: the serpentine marble has a much greater reflectance than hornblende schist so it is much paler in colour. The marble shows an Fe^{2+} absorption band at 1.0 μm, a water absorption band at 2.2 μm and a carbonate absorption band at 2.3 μm. The absorption band at about 2.0 μm is probably caused by the presence of water and CO_3^{2-}. The hornblende schist shows a very broad Fe^{2+} absorption band at about 1.0 μm. The other absorption bands appear to be due to water.

(b) The weathered andesite has a much reduced reflectance. The iron and water absorption bands seen in the spectrum of the fresh andesite have been lost as a result of weathering.

(c) The rock in the sunlight would appear to be much paler.

(d) The moist rock would appear to be darker.

(e) The images are sharper. Lakes and rivers appear black.

(f) At night. During the day much of the heat from the Earth's surface would be reflected. At night the radiation of heat absorbed during the day or produced by respiration can be detected in the absence of a reflected component.

(g) Angle of sun; shade from clouds; changes in vegetation; variation in snow or moisture levels.

(h) Numerous factors including temperature, light intensity, precipitation, topography, soil or rock type, wind strength, degree of grazing or browsing, stage of plant succession. (When an area is colonized by plants, a sequence of plant communities develops which leads up to a final, climax community.)

(i) With 10% grass cover, the spectral characteristics of andesite and limestone are partly obscured and identification of the rocks is difficult. With 30% grass cover, the spectrum is dominated by grass though the fall off in reflectance at 2.3 µm in limestone is still evident. At 50% cover, the rock spectra are completely obscured.

For soil the effects are not as strong. At 30% grass cover, the Fe^{3+} absorption band at 0.85 µm can still be seen and the OH^- band at 2.2 µm remains distinct. At 60% cover grass dominates the spectrum.

The spectral curves are most strongly affected at the shorter wavelengths because of the steep rise in the reflectance of grass at 0.68 µm.

12.17 (a) See Fig. A12.17.

(b) As a flood control measure. The dams also provide hydroelectric power.

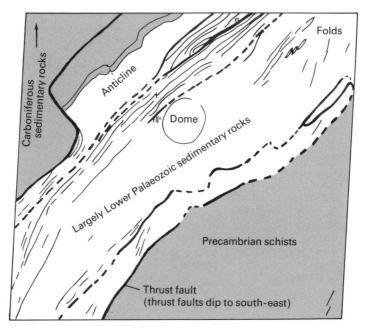

Fig. A12.17

12.18 (a) (1) Fact. (2) Hypothesis.
(3) Defined both as Avogadro's Hypothesis and Avogadro's Law. Started as a hypothesis, but seems to have become a law.
(4) Hypothesis. (5) Hypothesis. (6) Law of gravitation.
(7) Theory of plate tectonics. (8) Theory. (9) Hypothesis.
(10) Hypothesis. (11) Hypothesis. (12) Hypothesis.
(13) Hypothesis. (14) Hypothesis. (15) Law of constancy of interfacial angles. (16) Hypothesis. (17) Hypothesis.
(18) Fact. (19) Hypothesis. (20) Usually called a theory, but sometimes called a hypothesis. (21) Hilt's rule.
(22) Law of superposition. (23) Principle of uniformitarianism.
(24) Lasky's Law. (25) Law of faunal assemblage.

(26) Principle of isostasy.   (27) Law of included fragments.
(28) Law of original horizontality.   (29) Theory of evolution by natural selection.   (30) Snell's law.

(b) (i) 1 Models may be used to construct theories or to test and modify existing theories.

2 Models may be used for predicting or forecasting the effects of changing conditions on various systems (e.g. effects of burning fossil fuels on atmospheric temperatures; effects of population growth on resource usage).

3 Models may be used to simplify that which is very complex. For example, the way in which landscapes change through time has been described by the theory of the cycle of erosion. Landscape evolution is probably too complex to be described in such a simple way and it has been said that the cycle of erosion theory is really just a model.

(ii) 1 Initial reserves may not be accurately known. Potential reserves may become profitable at a late stage in a production cycle.

2 Changing prices, development and production costs may modify extraction rates. Improved technology may revive falling production rates or it may sustain high rates for longer than expected.

(c) (i) For both models, it is assumed that reserves are accurately known. In model 1 (the medium price scenario) it is assumed that the oil price will be 18 dollars a barrel till 1990, then it will rise to reach 43 dollars a barrel by 2000. In model 2 (the low price scenario) it is assumed that the oil price will be 18 dollars a barrel till 1990, then it will rise by 1% a year. Model 1 is thought to be the more probable.

(ii) There is a slower decline in production in model 1. Since this model uses the higher price for oil, it suggests that it will be worthwhile to produce from 46 probable and possible fields. With a fair price for oil, too, fields which are near exhaustion and which are producing only small quantities can remain in production.

(iii) They are similar in that both show a very rapid increase in production followed by a slower decline. However, the fall in North Sea production was not caused by a decline in reserves. It was caused by a rapid drop in the price of oil.

(iv) Production would remain higher for longer because even very marginal fields would be developed.

(v) By this time much of the major engineering work had been done (e.g. laying pipelines, building rigs) and the main phase of exploration had been completed.